Democracy as the Political Empowerment of the Citizen

Democracy as the Political Empowerment of the Citizen

Direct-Deliberative e-Democracy

Majid Behrouzi

LEXINGTON BOOKS

A division of
ROWMAN & LITTLEFIELD PUBLISHERS, INC.
Lanham • Boulder • New York • Toronto • Plymouth, UK

LEXINGTON BOOKS

A division of Rowman & Littlefield Publishers, Inc.
A wholly owned subsidiary of The Rowman & Littlefield Publishing Group, Inc.
4501 Forbes Boulevard, Suite 200
Lanham, MD 20706

Estover Road
Plymouth PL6 7PY
United Kingdom

First paperback edition 2006

British Library Cataloguing in Publication Information Available

The hardback edition of this book was previously cataloged by the Library of Congress as follows:

Behrouzi, Majid, 1956–
Democracy as the political empowerment of the citizen : direct-deliberative e-democracy.
p. cm.
Includes bibliographical references and index.
1. Democracy. 2. Information technology—Political aspects. 3. Information society—Political aspects. I. Title.
JC261.F68P53 2005
320'01—dc22 2005001173

ISBN-13: 978-0-7391-1028-7 (cloth : alk. paper)
ISBN-10: 0-7391-1028-4 (cloth : alk. paper)
ISBN-13: 978-0-7391-1809-2 (pbk. : alk. paper)
ISBN-10: 0-7391-1809-9 (pbk. : alk. paper)

Printed in the United States of America

™ The paper used in this publication meets the minimum requirements of American National Standard for Information Sciences—Permanence of Paper for Printed Library Materials, ANSI/NISO Z39.48–1992.

To my wife Elizabeth

Democracy, in a word, is a social, that is to say, an ethical conception . . . Democracy is a form of government only because it is a form of moral and spiritual association.

Personal responsibility, individual initiation, these are the notes of democracy. . . . There is an individualism in democracy . . . it is an ethical individualism; it is an individualism of freedom, of responsibility, of initiative to and for the ethical ideal.

—John Dewey (1888)

Democracy is the resolved mystery of all constitutions. Here the constitution . . . is returned to its real ground, actual man, the actual people . . . [here] . . . constitution appears as...the free product of men.

Democracy is *human existence*, while in the other political forms man has only *legal* existence. That is the fundamental difference of democracy.

—Karl Marx (1859)

[T]he cure for the ills of democracy is more democracy.

—Jane Addams (1902)

Contents

Preface

Democracy as the Political Empowerment of the Citizen: Direct-Deliberative e-Democracy conceptualizes the age-old idea of democracy in a new way. The fundamental idea underlying this new conceptualization is the now-neglected notion of the people's sovereignty. Literally, "democracy" means rule by the people. However, the people cannot rule unless they are empowered to do so. Since its inception, liberal democracy has eschewed the question of the people's sovereignty and their political empowerment for a variety of reasons. Among these, one should include the lack of faith of liberal democracy in ordinary citizens' ability to rule and the unavailability of practicable instruments and ways of empowering private citizens to act as sovereigns. Thus, liberal democracy's solution to the question of democracy has been the purely representative form of government that keeps citizens at a "safe" distance from the business of governing. The conceptualization attempted in this book resurrects the Rousseauean notion that the question of democracy is, not ultimately but, immediately the question of people's sovereignty. Moreover, this conceptualization relentlessly pursues the Rousseauean claim that sovereignty cannot be represented, and in order to be substantive, it ought to be exercised directly, hence *direct democracy*. The book takes the existing theoretical framework of American liberal democracy as its theoretical grounds and argues that the conception of democracy it develops is relevant to this society. In order to introduce the notion of sovereignty (and its direct exercise) into the liberal-democratic conceptual scheme, the book attempts to "individuate" the idea of the people's sovereignty via individuating the notion of the political empowerment of the people. That is to say, it conceptualizes the legislative power of the people as a composition that is made up of the sum total of the equal sovereign powers of the equal individuals who comprise the people or the nation. Of course, such a conceptualization would be meaningful only if there exist some feasible instruments or media that would empower individual citizens to exercise their individuated sovereign powers directly. The book argues that the present-day American society has such instruments and means at its disposal; i.e., it has both the material and technological means and infrastructures ("*e*-technologies"), and the political-cultural institutions needed for the actualization of the idea of the direct exercise of the individuated sovereign powers. In order to make a case for the practicability of the idea of the direct exercise of sovereignty by individual citizens, the book proposes and discusses a realistic democratic utopia wherein individual citizens are empowered to "fully express" their political wills in a direct manner via the use of *e*-technologies. These expressed wills are then "fully integrated" into a collective decision-making process that uses complex amalgamation schemes to "compose" collective wills on major legislative and policy questions.

In one sense, the conceptualization attempted here is a bold, and perhaps a last ditch, attempt to keep the idea of direct democracy (and its moral content) relevant to the contemporary world. This volume sets out to restore to democracy the ideal of citizens' direct participation in the democratic process, which its companion volume retrieves and reclaims. Concurrently published as *Democracy as the Political Empowerment of the People: The Betrayal of an Ideal*, the companion volume rescues the ideal of citizens' *direct* participation in decision-making from the perversions and distortions it has suffered throughout history. This ideal, it argues, was an essential component of the original idea of democracy. In this endeavor, both volumes take the insight they have derived from C. B. Macpherson's project of retrieving the moral content of democracy as a source of inspiration, and as their principal guiding thread.

The task of restoring to democracy its betrayed ideal takes the form of developing a new theory of democracy, which the book refers to as the "theory of direct-deliberative *e*-democracy." To this end, the theory develops a conception of individuated sovereignty with an eye on having it exercised directly by individual citizens through the use of *e*-technologies. This attempt also manifests itself in the importance the theory assigns to the idea of engaging everyday citizens in public deliberations on major legislative issues and public policy questions—deliberations taking place in both face-to-face and virtual forums using the *e*-media. In the idea of public deliberation, the theory of direct-deliberative *e*-democracy identifies two sorts of democratic utility. First, public deliberations motivate citizens to educate themselves on public issues, and provide directions for their civic-political self-educating activities. Second, public deliberations also help with generating "social capital" and strengthening communal bonds.

The theory of direct-deliberative *e*-democracy is highly critical of the liberal-democratic conception of democracy, the theories of deliberative democracy, and the received view on the question of "*e*-democracy." These approaches to the question of democracy either completely abandon the moral content of the idea by reducing it to a value-free method for selecting the government officials (the case of liberal democracy), or severely dilute its moral substance by overlooking the indispensability of the directness of citizens' participation to the idea of democracy (e.g., the case of deliberative democracy). For the theory of direct-deliberative *e*-democracy, the question of democracy (and democratic legitimacy for that matter) is not primarily about giving to people the "freedom of choice" in politics or the "right to choose" their governments; nor just about securing their consent; nor just about establishing "procedures" and assuring their "fairness"; nor just about establishing communication links between the constituents and their representatives (a main tenet of the received view of *e*-democracy). Nor is it just about morally justifying the power of authority and its right to exercise this power. Neither is this legitimacy primarily about the "quality" or "substance" of the laws legislated and the policies instituted; but also and primarily about the *actual, direct, and continuous input* of individual citizens into the legislative and policy decision-making process. More than anything, democracy is primarily about individual citizens being empowered to experience the political power directly, and to do so on a continuous basis. For the theory of direct-deliberative *e*-democracy, this constitutes both the immediate and ultimate yardstick of democratic legitimacy.

This work would not have been possible without the help and support of many individuals. I would like to express my gratitude to Professors Lesley Jacobs, Esteve Morera, and David McNally of York University (Canada), who directed the earlier stages of the development of this project and its companion volume. I am also indebted to Professor Richard Wellen of York University and Professor Richard Vernon of Western Ontario University for reading the manuscript and offering criticisms and suggestions. I also would like to express my gratitude to Professor Claudio Duran of York University, who directed me to read C. B. Macpherson. Special thanks go to Professor Joseph DeMarco of Cleveland State University, who encouraged and guided me throughout the entire project.

In addition, I would like to express my gratitude to Mr. Kevin Kay for reading parts of the manuscript and offering me helpful suggestions. Special thanks also go to Ms. Phyllis O'Linn, who read the entire manuscript several times and offered valuable suggestions and help. I am also thankful to Mr. Dick Wood for giving me permission to use his photo on the front cover of the book.

Lastly, I am greatly indebted to my wife Elizabeth, my family, and my friends who patiently endured me through this project and offered me love, encouragement, and support.

Introduction

Democracy as the Political Empowerment of the Citizen: Direct-Deliberative e-Democracy is being published concurrently with its companion volume *Democracy as the Political Empowerment of the People: The Betrayal of an Ideal*. The present volume starts where the companion volume leaves off. *The Betrayal of an Ideal* offers a critical examination of the hitherto-existing theories and regimes of democracy. The primary aim of this examination in *The Betrayal of an Ideal* is to retrieve the ideal of citizens' *direct* participation in the political process and lay the political-theoretical grounds for reclaiming the ideal, which constitutes the subject matter of the present volume. *The Betrayal of an Ideal* sets the stage for arguing that the notion that citizens should have direct and substantive roles in making legislative and policy decisions is a deep-seated idea in the Western tradition of political thought, and constitutes the main ideal (and a primary political-moral component of the idea) of democracy.[1] It further argues that, this ideal (and its moral content) has been sabotaged time and again as the idea of democracy has been subjected to perversion after perversion throughout history. Moreover, *The Betrayal of an Ideal* argues that, with its system of political representation, the "liberal-democratic conception of democracy" represents the latest, and most egregious, version of these perversions.[2] Finally, it argues in passing that present-day society has all of the material, technological, social-political, and cultural prerequisites necessary for reviving the "original" idea of democracy.

Building upon what *The Betrayal of an Ideal* retrieves, the present volume sets out to formulate a new theory that restores to democracy its ideal of the citizens' direct participation in legislative and political decision-making. As part and parcel of formulating the new theory, this volume develops a philosophical foundation for the now-ubiquitous idea of "*e*-democracy," and uses this foundation to argue that the idea merits serious consideration. The relevance of the idea of *e*-democracy to the task at hand, as this volume argues, lies in that the idea is capable of transforming our understanding of what it means to have citizens directly participate in the political process in the modern day nation-states, especially in the United States.

The latest innovations in electronic and communications technologies ("*e*-technologies") in recent decades, and the ever-growing trend in using their immense powers in all aspects of economic activities and social interactions have given rise to the idea of utilizing these technologies in the service of democracy. In recent years, many enthusiasts of this idea have thought of numerous innovative schemes and methods that could realize this idea. These schemes range from the idea of holding "electronic town meetings" for the purpose of public deliberations and consensus-building to having citizens vote *directly* online on public issues at both the local and national levels.[3] A good example of voting online is the experiment that took place in the state of Arizona in the United States on March 7, 2000. In the presidential primary election in this state, the Democratic Party held the world's first legally binding electronic vote using the Internet.[4] The Democratic Party repeated the experiment with greater success in Michigan's presidential primary in February 7, 2004.

For those enthusiasts who interpret democracy literally as the idea of direct self-rule by the people, especially for those who equate the people's direct self-governance with ancient Athens' short-lived experience with direct democracy, the new innovations in the electronic and communications technologies are dreams come true. Thanks to these innovations, ancient Athens' political universe can now be reconstructed in the virtual plane, and its model practiced as a viable alternative to today's purely representative form of democracy.[5] The latest innovations in *e*-technologies have made it possible for these enthusiasts to imagine a new form of political universe, a *direct democracy* indeed, where the people themselves have turned into the actual decision-makers, and the professional politicians into the servants of the people in the true sense of the word.

Predisposed to the view that the current fascination with the democratic implications of the new electronic and communication technologies is more than a passing fad and that these technologies have the potential to transform present-day American society's understanding and practice of democracy, this volume sets out to put the question of what is often referred to as "*e*-democracy" in a philosophical context and use it as a foundation for developing a new theory of democracy that shines the spotlight on the citizens' direct and deliberative participation in the political process. Toward this end, the present volume argues that the new *e*-technologies and *e*-media not only have provided the impetus for revisiting the original idea of democracy and retrieving the value inherent in the idea of the citizens' *direct* participation in politics, but also that they have made it possible to reformulate the idea of direct democracy in ways that would make it a viable option and worthy of serious consideration as an alternative approach to the question of democracy in today's large nation-states.

In formulating this theory, the present volume attempts to conceptualize the idea of the citizens' direct participation in decision-making in a new way. In sharp contrast to the traditional understanding of the concept that treats democracy as the "political empowerment of the people," this volume conceptualizes democracy as an idea that takes the *political empowerment of the citizen* as its primary subject matter. The main problem with conceptualizing democracy as the political empowerment of the people (or "power to the people") is that it leaves democracy vulnerable to elitist subversions. That is to say, it allows some to argue that it is possible—or actually desirable—for a select group of individuals (whether the elected representatives of the people or their self-appointed guardians) to function as the instruments of the political empowerment of the people. (As *The Betrayal of an Ideal* shows, the elitist subversion is one of the two main senses of the perversion of the idea of democracy.) Conceptualizing democracy as the political empowerment of the individual citizens, the book argues, guards against this defect suffered by the prevailing understanding of democracy.

In conceptualizing democracy as the political empowerment of the individual citizens, the question of empowering the people is then transformed into the question of equally empowering the equal individual citizens who comprise the people. This approach to the question of democracy resurrects the Rousseauean ideas that democracy is about the popular sovereignty and that popular sovereignty cannot be represented. The book relentlessly pursues these ideas and attempts to incorporate them into the existing theoretical framework of the American liberal democracy. This incorporation requires that the sovereign power of the people be "individuated" into the sovereign powers of the individual citizens.[6] Thus, the question of the exercise of sovereignty by the people becomes the question of the individual citizens exercising their individuated sovereign powers directly. The direct exercise of this power is conceptualized as the direct participation of the individual citizens in making major policy and legislative decisions.

The conception of "democracy as the political empowerment of the citizens" is expressed in terms of two core principles: "the macro principle of the political sovereignty of the individual" and "the micro principle of the 'social autonomy' of the individual." The former principle maintains that political society is to be organized in such a way that the individuals (the citizens) would be empowered to exercise their sovereignty *directly*, and to do so on an *ongoing basis* at macro levels—that is, via *fully expressing their political "wills," and directly incorporating these expressed wills into collective political decision-making processes* at the national, state, and local levels.[7] The latter principle, on the other hand, carves out a space for the individual at the micro levels of workplace and community and thus enables her to exercise some degree of autonomy in these social units. (The focus of the book is on the macro principle.) At the heart of the notion of having individuals exercise their sovereign powers and social autonomy in a direct manner lies a multi-layered scheme of electronically-facilitated voting that enables each and every individual to both "fully express" her positions and wills on issues, and to have these expressions "fully incorporated" into the decision-making processes.

On numerous occasions, the book argues that its "conception of democracy as the political empowerment of the citizen" is superior to the "liberal-democratic conception of democracy" which theoretically and constitutionally bars the ordinary citizens' from having a direct and meaningful participation in the social decision-making process. On some other occasions, the book argues that its conception of democracy is also superior to the deliberative conceptions which take democracy "as a matter of discourse" and not primarily as a matter of action or direct participation. Deliberative democracy limits the people's roles in politics to making *potential and indirect contributions* to decision-making, thus falling short of addressing the question of citizens' *actual and direct participation* in the political sphere. In the deliberative scheme of things, citizens' deliberations produce "agreed judgements" and "collective verdicts." However, these judgements and verdicts do not have the status of laws but only that of suggestions, messages, and perhaps mandates that would be delivered to the lawmakers (the political elite, the representatives) to be considered by them at their discretion in making policy or legislative decisions. Thus conceived, deliberative democracy is nothing but the *idealized form* of a pure representative democracy. Theories of deliberative democracy do not and cannot address the question of the people's sovereignty posed by Rousseau. Rousseau's criticism of representation as a form of slavery remains relevant to the theories of deliberative democracy.[8] The book further argues that unlike the theories of participatory and deliberative democracy which attempt to give substance to (and thus strengthen) the democratic component of liberal democracy via weakening its liberal component, the conception of democracy advanced here strengthens both components by giving real substance to both of them.[9]

The driving force behind conceptualizing "democracy as the political empowerment of the citizen" in this book is the conviction that the question of democracy or democratic legitimacy is—not ultimately, but *immediately* and *directly*—the question of citizens' political sovereignty. Moreover, democracy or democratic legitimacy is not primarily about giving to people the "freedom of choice" in politics or the "right to choose" their governments (liberal democracy). Nor is it just about securing the "consent" of the people, nor just about establishing "procedures" (liberal democracy) or assuring their "fairness" (deliberative democracy). Nor is this legitimacy just about morally justifying the right or power of authority to make laws (liberal democracy, deliberative democracy); nor just about the "output"—i.e., the "content of outcomes" or "substance"—of decisions made by the authority (participatory democracy, deliberative democracy); but also, and primarily, about the *direct and continuous input* of citizens into the decision-making process. *More than anything, democracy is primarily about individual citizens experiencing political power directly and doing so on an ongoing basis.* Consequently, the yardstick of democratic legitimacy is the degree to which this ideal is realized. Democracy is primarily about providing and facilitating the highest feasible degree of the actual—i.e., *direct* and *ongoing*—participation by citizens in legislating the fundamental laws they abide by, and in making decisions about the fundamental policies that affect their lives.

As a way of illustrating how this conceptualization of democracy could work in practice, the book constructs a "Realistic Democratic Utopia." The theoretical-institutional framework of the Realistic Democratic Utopia produces a set of institutional arrangements that would work cooperatively with the existing institutions of liberal democracy to provide a higher level of democracy than presently available. The institutional arrangements in question would have technological structures that would rely heavily on the latest innovations in the information and communication technologies in order to facilitate citizens' direct participation in making macro decisions. These institutional arrangements would make it possible for citizens to participate both in public deliberations *and* in actual decision-making on major national issues via electronic town hall meetings and electronic voting schemes. Given that public deliberations would be an integral part of this conception, and also given that these institutions would be electronically networked and that voting would take place via electronic media, the theory of democracy developed in this book is referred to as the "theory of direct-deliberative *e*-democracy."

The book offers a two-track justification for the theory of direct-deliberative *e*-democracy. On the one track, it puts forth the theory as capturing the main ideal of the original idea of democracy, viz., that of citizens' direct participation (and thus remaining true to its original meaning), and offers it as an option that is both viable and worthy of serious consideration as an alternative approach to the question of democracy in today's large and complex nation-state. On the other track, the book presents the theory of direct-deliberative *e*-democracy as a serious contender on the strength of its own justificatory arguments: its developmental argument, and its attempt to provide an essentially proceduralist argument for democracy without losing sight of democracy's substantive and epistemic dimensions. In its developmental argument, the theory links the value it affirms in the idea of the citizens' direct participation to the ideal of the development of the human individual in ways similar to the arguments presented by J. S. Mill, Karl Marx, John Dewey, and C. B. Macpherson. Once the theory of direct-deliberative *e*-democracy is fully elaborated, the book proceeds to argue that the theory has the potential to solve, or dissolve, some of the long-standing questions that have dogged democratic theory throughout its history.

* * * * *

The formulation of the theory of direct-deliberative *e*-democracy will take place in stages. As the first stage in this endeavor, Chapter 1 will conceptualize democracy as the *political empowerment of the citizen*. This formulation will take place within the confines of the liberal-democratic framework. However, rather than trying to strengthen the democratic pole of the liberal-democratic conception by weakening its liberal pole—which was the strategy of the theories of participatory democracy—this new formulation will attempt to strengthen the democratic component of this conception by breathing substance into its liberal pole. Alternately stated, this approach will attempt to revive the pre-liberal content of the idea of democracy and integrate it into the theoretical fabric of present-day American liberal democracy.

The conceptualization of the idea of democracy as the political empowerment of the citizen will first be presented in the language of "classical" democratic theory, mainly in Rousseau's, and then later will be brought into conformity with the current language of liberal democracy. This conceptualization will be expressed in terms of two *ideal* principles of the *political sovereignty* and the *social autonomy* of the individual. In *concrete* terms, the idea of the political empowerment of the citizen will be postulated as the power to "*fully express*" one's political wills and the power to "*fully integrate*" these expressions into the collective policy- and decision-making processes at various levels of society.

As the second stage of formulating the new theory, Chapter 2 will present a discussion of how these two powers can be actualized in present-day American society. Here, the discussion will mainly center on devising a complex collective decision-making scheme that will be referred to as FEFI—("Full Expression and Full Integration"). The process of developing the scheme in question begins with presenting first a simple system of voting that would give voters virtually infinitely many different ways of expressing their wills and then would incorporate these expressions as inputs into an amalgamation-composition process that would produce a "collective will" or a collective decision. This presentation is then followed by a discussion of a number of potential criticisms that could be directed against the scheme, including the charges that the resulting collective wills would be "inaccurate" and "unfair" expressions of the "will of the people." Based on this discussion, the scheme will then be revised to produce a more complex system that would be able to deflect most criticisms.

The third stage of formulating the theory of direct-deliberative *e*-democracy will take place in Chapter 3. This will be done by constructing what will be referred to as the "Realistic Democratic Utopia" that will serve as a theoretical framework within which one can attempt to develop a practical model for the actualization of the conception of democracy developed in Chapter 1. The Realistic Democratic Utopia will employ the complex voting scheme developed in Chapter 2 as its primary collective decision-making mechanism. Here it will be argued that present-day society has the necessary technological infrastructure and other relevant material prerequisites for realizing the idea of the Realistic Democratic Utopia.

The basic model for citizens' direct participation in collective decision-making in the Realistic Democratic Utopia would be as follows: before a vote on a "major" bill, issue, or policy takes place, there would be ample discussions and informative sessions that would help citizens educate themselves on the issues for a specified number of weeks or months. These sessions would be directed primarily by the "guardians of the public interests" (the elected "experts" and "trustees" of the people) both in person (in the local community and town meetings) and in the media (including the "electronic town meetings" and various television channels and radio stations publicly funded and operated solely for this purpose). During these sessions or in their aftermath, citizens would debate and deliberate both on the virtual plane (whether one-to-one or through "electronic town meetings" and online discussion forums, and in the mass media) and in the actual world in public places (e.g., the workplace, local community meetings, "civic homes," "local talk shops," and social gatherings).[10] Then, at a specified day and time block, the citizens would vote by using their electronic voting cards and PINs in conveniently-located and secured voting precincts (or possibly on the Internet). The elected policy "experts" and the "trustees" of the people, as well as the members of the Congress, would also vote on the issue in their respective assemblies in the same time and day block. The aggregated votes of citizens would be weighed against the aggregated votes of the "guardians," representatives, and senators in accordance with a set of carefully developed formulas. Higher popular participation in voting would assign higher weight to the votes of the citizens, in comparison to the votes of the guardians and representatives. Moreover, the voting scheme employed would not be the existing monosyllabic yes or no. Rather, it would be a multi-layered scheme that would allow each and every individual to "fully express" her views and "wills" (both private and public), as well as having these expressions "fully incorporated" into the decision-making process.[11]

As the fourth and final stage in formulating the theory of direct-deliberative *e*-democracy, Chapter 4 brings together the "conception of democracy as the political empowerment of the citizen" (developed in Chapter 1), the voting scheme of Chapter 2, and the theoretical-institutional framework of the Realistic Democratic Utopia (explored in Chapter 3), and synthesizes them into a single theory. This synthesis takes place against the backdrop of the "liberal-democratic conception of democracy." The main political value affirmed by the "theory of direct-deliberative *e*-democracy" is the value inherent in the principle of the citizens' direct participation in the legislation of the major laws by which they abide. The second political value affirmed by the theory of direct-deliberative *e*-democracy is the principle that social-political decision-making ought to be embedded in knowledge, moral understanding, and virtue.

It is argued in Chapter 4 that the theory of direct-deliberative *e*-democracy, first and foremost, is committed to the realization of the idea of the unencumbered and fullest feasible (positive) development of the human individual, which it regards as the ultimate value of the human universe and society's *raison d'être*. Moreover, it is argued that the theory is built around a faith in the abilities of ordinary citizens to make sound political decisions. Furthermore, the hallmark of the theory of direct-deliberative *e*-democracy is that it rests on a thick notion of sovereignty. The theory is relentless in the pursuit of the idea that democracy is primarily to be identified with the individual citizens' *actual, direct,* and *continuous* exercise of sovereignty (i.e., their *direct and continuous participation* in social decision-making on major issues). On this question, the theory diverges considerably from the liberal-democratic conception, which is premised on a limited, indirect, and intermittent-periodic exercise of sovereignty by citizens that takes place exclusively during the election of representatives. Another feature of the theory of direct-deliberative *e*-democracy is that its thick notion of sovereignty is coupled with a thin notion of equality in the realm of material holdings. Though this thin egalitarianism of the theory of direct-deliberative *e*-democracy goes against the grain of tradition in the "classical theory" of democracy, it is nevertheless a consequence of developing the theory within a liberal-democratic framework.

The Conclusion revisits some of the themes discussed in Chapters 1-4, and in *The Betrayal of an Ideal*, and further develops some of the arguments presented earlier. It argues that the ultimate guarantor of a responsible government—as well as the true remedy for the democratic shortcomings of liberal democracy—is a *government "by" the people* where individual citizens directly participate in performing some of the legislative and policy-making functions of governing. The Conclusion also suggests that some of the longstanding issues and problems in political theory and the theory of government can be solved or resolved by the theory of direct-deliberative *e*-democracy. Four such questions are treated in greater details than others. First, it is argued that direct-deliberative *e*-democracy resolves the dichotomy drawn by Benjamin Constant between the "liberty of the ancients" and the "liberty of moderns."[12] Second, it is argued that direct-deliberative *e*-democracy makes it possible for a democratic society to assimilate those aspects of Plato's guardianship theory of government that makes it attractive, viz., the idea that social-political rule ought to be embedded in knowledge, moral understanding, and virtue. Third, the Conclusion argues that the theory of direct-deliberative *e*-democracy can lay claim to offering a solution to a fundamental philosophical problem that hounded ancient and early modern political thought, namely, the now-forgotten problem of reconciling the idea of governing by the consent of the ruled with the equally desirable (yet elitist) idea of governing in accordance with principles of reason and wisdom. Finally, as an extension of the discussion of the first question, the Conclusion suggests that the theory of direct-deliberative *e*-democracy can be viewed in some respects as a unique synthesis that unifies, in a coherent fashion, the liberal theory of government (the idea of liberal constitution and liberal institutional arrangements) with the most compelling features of republican political thought (the ideas of public-spirited political culture and active citizenship).

Notes

1. In treating democracy as a normative concept, *The Betrayal of an Ideal* defined democracy as meaning "rule by the people," or as citizens having the *actual* and *direct* sovereign authority in the state. (This is the "original" definition of the idea.) Starting with this definition, *The Betrayal of an Ideal* postulated democracy in terms of a set of moral-political values that were associated with democracy in its original formulation. These values were expressed in terms of three *ideals* that *The Betrayal of an Ideal* regarded as having guided the theory and practice of ancient Athenians. These ideals were: (1) the ideal of the *direct participation* of citizens in legislative and policy decision-making, (2) the ideal of *substantive equality*, and (3) the ideal of *public deliberation. The Betrayal of an Ideal* privileged the ideal of direct participation as the "true" and "main" ideal of democracy—and its primary moral component—over the ideals of equality and deliberation for two reasons. First, this ideal needs to be the focus of attention for it has suffered the most sabotage in the past and, at the same time, has received the least attention in the recent literature devoted to the goal of retrieving the moral content of democracy. Second, the values inherent in the ideal of direct participation is directly linked to the values associated with the ideal of the full and positive development of the human individual which, in Chapter 1 of this volume, will be privileged as the ultimate value of the human universe. See Chapter 1 of *The Betrayal of an Ideal* for further details.

2. The "liberal-democratic conception of democracy" was introduced in Part II of *The Betrayal of an Ideal.* Briefly stated, this conception regards democracy as the idea of the rule by a freely and popularly elected representative government. Part II faulted this conception for being elitist. It also took issue with the purely representative form of government prescribed by this conception as it argued that this form of government is more

attuned to the interests of the propertied classes, and is ultimately tantamount to political disempowerment of the people. Part III of *The Betrayal of an Ideal* went a step further and characterized the "liberal-democratic conception of democracy" in its most recent manifestation as "audience democracy" and "fund democracy."

3. A survey of some of these ideas is provided in the opening section of Chapter 3.

4. A privately owned company named Election.Com conducted this election.

5. For instance, Rheingold speaks of "Athens without slaves," and expresses the optimism that "computer-mediated communications" (CMC) could "revitalize citizen-based democracy" and could "have democratizing potential in the way that alphabets and printing presses had democratizing potentials" (Rheingold 1993, p.14 and p.279, respectively).

6. As will be seen in Chapter 1, the notion of the "individuating" sovereignty does not contradict Rousseau's claim that sovereignty is "indivisible."

7. It should be noted at the outset that throughout this work, the term "will" is used in a technical sense. A "will," as will be discussed in detail in Chapter 1, denotes a "public judgement"; that is to say, a particular sort of opinion on a given issue that; a) is well-informed and educated on the issue and has the knowledge of alternatives, b) is subjected to reflection on the basis of moral considerations and reasonableness, c) is preferably scrutinized in public deliberations, and d) has the "force of the commitment" of the individual citizen who holds it. As will be seen in Chapter 1, an individual citizen may have more than one "will" on any given public issue.

8. Theories of deliberative democracy are discussed in Chapter 13 of *The Betrayal of an Ideal.* Concisely stated, the fundamental idea defining theories of deliberative democracy is the contention that democracy derives its legitimacy from the participation of the citizens in public *deliberations* on the issues of concern to the society. In the words of Joshua Cohen, one of the original proponents of the idea, "[t]he notion of a deliberative democracy is rooted in the intuitive ideal of a democratic association in which the justification of the terms and conditions of association proceeds throughout public argument and reasoning among equal citizens. Citizens in such an order share a commitment to the resolution of problems of collective choice through public reasoning, and regard their basic institutions as legitimate insofar as they establish the framework for free public deliberation" (Cohen 1997a, p.72). *The Betrayal of an Ideal* faulted theories of deliberative democracy for ignoring the democratic ideal of *direct* participation, and thus for their inability to overcome the two glaring democratic shortcomings of liberal-democracy, which it characterized as "audience democracy" and "fund democracy."

9. Theories of participatory democracy are discussed in Chapter 12 of *The Betrayal of an Ideal.* Briefly stated, theories of participatory democracy came out of the movements of the 1960s and were keen on emphasizing the moral substance of democracy, as well as emphasizing the importance of citizens' participation in the public life—at the level of community (the version promoted by the "Students for a Democratic Society"), at the workplace (Carole Pateman's version), and in the political decision-making about economic life (C. B. Macpherson's version). While Pateman emphasized the education utility and community-bonding effects of democratic participation, Macpherson focused primarily on retrieving the moral content of democracy, as he posited democracy as "a set of moral ends." Macpherson regarded this content primarily as *substantive* (i.e., economic) equality—to be contrasted with the *formal* interpretation of equality in liberal-democracy. For Macpherson, the project of retrieving the moral content of democracy was closely linked to the goal of dethroning the "market morality" and market-based conceptions of

Man, and thus ultimately to reclaiming Man as an ethical being and a developer of his human capacities.

10. The phrases "civic homes" and "local talk shops" are borrowed from Barber (1984), p.271 and p.268. The main idea behind these institutions is discussed in Chapter 13 of *The Betrayal of an Ideal*, and will be further discussed in Chapters 3 and 4 in this volume.

11. There would be about 10-12 "major" issues to be decided annually (an average of 3 issues in each voting occasion, scheduled 4 times a year). Examples of major issues would be the national education policy and annual budget planning. On each major issue, voters would choose among a meaningful range of options by a scheme that will give them virtually infinitely many different ways of expressing their views. They would also have the "protest option" in case they are dissatisfied with the range or meaningfulness of the options presented to them on the ballot. The protest option can also function as a "veto vote" if it draws a specified percentage of the votes. A "veto vote" is an indication that voters did not approve of the policy options presented to them and the agenda-setters must put together another set of policy options based on studying the results and then schedule another public voting session.

12. According to Constant, the ancients exercised direct sovereignty while suffering from the lack of negative civil liberties. On the other hand, the moderns, living in liberal societies, have negative civil liberties while being deprived of exercising sovereignty directly. Constant—in his address to the *Athenee Royale* in 1819—argued that it is not possible to combine direct sovereignty with civil liberties. See Constant (1988), pp.309-28 for the full text of his speech.

Chapter 1

Conceptualizing Democracy as the Idea of the Political Empowerment of the Citizen

The Betrayal of an Ideal presented a critical examination of the main conceptions of democracy that had appeared on the political horizons of Western civilization since the inception of the idea. The presentation began with an appraisal of the experience of ancient Athens and ended with the examination of the "theories of deliberative democracy" at the turn of the millennium.[1] The main theme of *The Betrayal of an Ideal* was the contention that these conceptions, in various shades and forms, had distorted the original meaning of the idea of democracy. In discussing the more recent conceptions, *The Betrayal of an Ideal* also argued that for a variety of reasons, these conceptions had either failed to produce institutional designs and models for practicing the idea in its true sense, or proved unsuccessful in practicing these models in sustained manners, in case they had managed to produce them.

Two commitments guided, as well as motivated, *The Betrayal of an Ideal*. The first is the devotion to the idea that one needs to cut through the prevailing deception and bring to light the now-neglected fact that the idea of the citizens' *direct* participation in decision-making not only represents the true sense and the original meaning of the idea of democracy, but more importantly, it constitutes the main *ideal* (and the fundamental *moral* component) inherent in the idea.[2] The second, on the other hand, is the conviction that this ideal ought to be rescued from the distortions it has suffered throughout the life of the idea, and be reclaimed as the hallmark of democracy in the contemporary world.[3] This latter commitment, which made its presence felt only indirectly in *The Betrayal of an Ideal*, will serve as the guiding principle in this volume. In reclaiming for democracy the ideal of the citizen's direct participation in decision-making, starting with this chapter, the present volume will set out to formulate a new conception of democracy that will lay claim to both remaining true to the original meaning of the idea and being able to produce an actual model and institutional design for practicing it.

Holding on steadfastly to the idea that democracy is primarily about demos being sovereign and exercising their sovereignty *directly*, the new conception of democracy to be formulated in this volume will take an entirely different approach to the question of democracy. As will be seen, the approach here will remain squarely within the confines of the liberal-democratic theoretical framework. However, rather than trying to strengthen the democratic pole of liberal-democracy by weakening its liberal pole, as was the strategy of the "theorists of participatory democracy," the approach here will attempt to strengthen the democratic component by breathing substance into the liberal pole of the liberal-democratic conception.[4] As will be seen, the approach will attempt to revive the pre-liberal content of the idea of democracy and integrate it into the theoretical fabric of American liberal-democracy. Compared with the "liberal-democratic conception of democracy," what will be markedly different in this new conception is that the existing representational institutions in the liberal democracy will be supplemented with a new set of institutions that would empower citizens to participate *directly* in the political decision-making process.[5]

Toward presenting this new conception of democracy, this chapter will attempt to conceptualize the idea of democracy, not as the political empowerment of the people ("power to the people"), as has been done traditionally, but as *the political empowerment of the individual citizens* (*"power to the individual citizens"*).

Starting with this chapter, the terms "demos" and "the people" will be used to refer to the *entire* population throughout the book. Moreover, they will also be taken to mean nothing more than the sum total of the politically and legally equal individuals (citizens) who comprise the population. The main implication of these assumptions is that they make it possible to treat the political empowerment of the people or demos as the *sum total of the political empowerments of the individual citizens comprising the demos*. This new formulation will be referred to as the "*Conception of Democracy as the Political Empowerment of the Citizen*" in the rest of the book. Moreover, as will be seen shortly in this chapter, the new formulation will bring forth, and join together, two old concepts of political philosophy, viz., *sovereignty* and *autonomy*. In this new reformulation, the concept of sovereignty—which often escapes the attention of contemporary political philosophy—will carry its traditional meaning, viz., *the power of the people to legislate laws*; and as such will be treated as a principle that is applicable and operative mainly at the "macro" levels or large-scale political arenas such as the state/local levels and federal/national politics.[6] Autonomy, on the other hand, will keep its usual meaning, viz., the individual's self-direction, but will be employed in a different sense than that in which it has traditionally been used. In its traditional sense, autonomy is a concept that is relevant only to the private or moral domains of the individual's life. Here, by contrast, autonomy will be treated as a concept applicable to small public spheres, or the "micro" or local levels of the individual's life, e.g., the workplace. To distinguish the sense of autonomy intended here from its traditional sense, the term will be used in conjunction with the word "social" to produce the phrase "*social autonomy*." As conceived, social autonomy will come to designate the power of the individual to exert control over, or give direction to, those social-political elements that play important roles in shaping or affecting her everyday life, such as the ones present at her workplace.

Finally, it must be stressed at the outset that the conceptualization of the idea of "democracy as the political empowerment of the citizen" is not intended as a transhistorical or general definition of the concept, but only as one offered against the backdrop of present-day American society. As will be seen, the theory of direct-deliberative *e*-democracy requires certain levels of socio-economic developments, highly developed political-legal structures, and certain degrees of material-technological achievements, all of which, this book assumes, have sufficiently been attained in present-day American society. Among these, one can list the following: a highly productive economy; a legal-political system that upholds and protects the civil and political liberties of the individuals, and safeguards them against the "tyranny of majority"; a "public political culture" that values the ideas of respect for persons, fairness, pluralism, and tolerance; and finally a vast system of electronic-technological infrastructures and capabilities.

Democracy as the Political Sovereignty and Social Autonomy of the Citizen

Having mentioned these preliminaries, the conceptualization of the new idea of democracy can now be offered: Democracy is to be treated as a concept that deals directly with the question of the political empowerment of the individual citizens. It is a concept that brings together and conjoins the traditional concepts of "(political) sovereignty" and "social autonomy." In other words, democracy can be formulated as a conception founded on the following two *ideal* principles:

1. *The ideal macro principle of political sovereignty of the individual (citizen)*: In an ideal democracy, the "political society" (the state) is organized in such ways that the individual citizen is empowered to exercise her sovereignty and do so *directly and on a continuous basis.* That is, in an ideal democratic society, the individual citizen would have substantial, substantive, and direct input in formulating the major public policies, and in making decisions and legislating the laws regarding major issues. (She could also have input in formulating and legislating minor policies and laws, if she chooses.)

2. *The ideal micro principle of the "social autonomy" of the individual (citizen)*: In an ideal democracy, "civil society" and "economic society" are organized in such ways that the individual is empowered to exercise her "social autonomy." That is, in an ideal democratic society, the individual would have substantive input in, and substantive control over, the

> social, economic, and political factors that play important roles in her everyday life activities at work and in the social units in which she participates.[7]

The formulation being attempted here is not intended as an essentialist definition of democracy, but as a maximal or optimal one that gives the fullest possible expression to the original idea of democracy. Nor is this formulation offered as a universal or transhistorical formulation of the original idea of democracy, but rather as a potential one for present-day American society. Without a doubt, the two principles underpinning this conceptualization are manifestly value-laden, and aim at recapturing the meaning and the moral content of the original idea of democracy. These principles explicitly affirm the value of direct participation. As will be seen in the following section, this value draws its content from more substantial value claims. However, the fact that this conceptualization is offered in terms of *ideals*, is an indication that this formulation espouses Frank Cunningham's treatment of the idea of democracy as a *quantitative* concept in his *Democratic Theory and Socialism*. That is to say, the "conception of democracy as the political empowerment of the citizen" regards democracy as an ideal or a qualitative state that can only be approximated, but may not be fully realized. In other words, the actualization of the idea of democracy is to be regarded as a "matter of degree," and not an attribute or quality that a social-political order or unit can either lack or possess.

In treating democracy quantitatively, it then follows that one should characterize social-political orders or institutions only in terms of less-democratic or more-democratic, and not in qualitative terms of democratic, non-democratic, or anti-democratic.[8] Moreover, given the quantitative treatment of the conception of democracy offered above, one should expect that the rise to higher levels of democracy, and thus fuller realization of the value of direct participation, could only be achieved gradually. Finally, it should be noted that, no doubt, the "liberal-democratic conception of democracy" and its current practice in present-day American society would score extremely low on the democratic scale if measured against the two principles outlined above. In what follows some of the essential features of the macro and micro principles will be discussed in passing. A more detailed examination of each of these principles will be provided later in this chapter.

The point of departure of the macro principle of political sovereignty of the individual, as sketched above, is a positive reading of Rousseau's conception of sovereignty. This particular reading takes sovereignty as a concept that stands for the collective exercise of the power of decision-making by the people; that is, as the idea that "the people that is subject to the laws ought to be their author."[9] What is new in the macro principle of (political) sovereignty of the individual is the attempt to *individualize*, or rather *individuate*, the Rousseauean conception. In doing so, the macro principle follows in the tracks of the "*methodological* atomic individualism"—an approach that takes the individual, "conceived atomistically," as "the point of departure for moral, social, and political theory."[10] The idea of individuating sovereignty begins with stripping the terms "the people" and "demos" of their customary sociological and group- or class-related properties, and thus with taking these terms to mean nothing more than the arithmetic sum of the politically and legally equal individual citizens comprising them. Conceived as such, the notion of the sovereignty of the people turns out to be *the sum total of the individuated sovereignties of the individual citizens who comprise the demos.*

It is important to note that positing "the people" or "demos," and consequently the individual citizens who comprise them, as free of class- or group-related attributes is not tantamount to denying that these attributes affect or condition (and in some cases in significant ways) many important spheres and aspects of life-activities in present-day society. What motivates this assumption is that it seems to be the only plausible approach to giving a concrete and real expression, as well as a real substance, to the idea of the political empowerment of the *entire* people. Moreover, one needs to note at this point that the idea of the "*individuation*" of sovereignty must not be mistaken with its "divisibility." The idea of individuating sovereignty does not contradict Rousseau's claim that sovereignty is "*indivisible*"; and thus it should not be confused with what Rousseau rejects. By claiming the indivisibility of sovereignty, Rousseau mainly meant to rebuke those political theorists who, in their advocacy of division of the state powers into legislative and executive, or into internal affairs and external affairs, lose precision in what they advocate, and confuse laws—(which are supposed to be general in object and source)—with their particular applications, e.g., a particular policy.[11]

Furthermore, it should be added that the notion of the individuation of sovereignty proposed above does not come into conflict with the Rousseauean principle of the indivisibility of sovereignty for another reason. That is to say, the notion in question does not create the situation wherein various loci of legislative power and authority compete with one another for political power, a scenario that would result in political instability—and without a doubt this is what Rousseau intended to avoid. Rather, what the notion of individuating sovereignty does is to spread the legislative power *evenly* and *equally* throughout society, and thus facilitate the *direct* participation of each and *every* individual in political decision-making. The goal of facilitating the direct participation of the individual citizens, it should be emphasized, is the main motivation for individuating the notion of sovereignty, and thus, the main reason for formulating the macro principle of the political sovereignty of the individual. Furthermore, it should be added that the idea of individuating sovereignty is also intended to both acknowledge the commitment of the American liberal-democracy to individualism (i.e., its commitment to the principles of the moral and social primacy of the individual), and to give substance to it by extending it to the realm of the political. Finally, another factor that motivates the idea of individuating sovereignty is the fact that present-day American liberal democracy, unlike the hitherto existing societies, has at its disposal the social-political and material-technological structures and means that could help transform the notion into a practical idea. (These capabilities of present-day American society will be examined in Chapters 3 and 4.)

In comparing the macro principle of political sovereignty of the individual with Rousseau's conception of sovereignty, one should also take note of the following difference. Rousseau assumes that the people enact the laws exclusively based on the consideration of their *common interests.* In contrast to Rousseau's, the main idea behind the macro principle of the sovereignty of the individual is that the people ought to be sovereign not as a *whole,* or as a *uniform or homogenous entity* (i.e., the nation or the overwhelmingly large and dominant class), but rather as individual citizens. As one might suspect, a direct consequence of individuating sovereignty would be to open the door for the individual's private interests to enter into the pool of interests that the notion of sovereignty encompasses. In other words, it is not just the common interests of a nation or a class that the macro principles of the political sovereignty of the individual would oversee. As will be seen later, this conception of sovereignty would also encompass the private interests of each and every individual, but only to the extent that it would be feasible for individual citizens and society to integrate the considerations of private interests into the formulation and legislation of policies and laws.

Having outlined the essential features of the macro principle of the sovereignty of the individual, a concise discussion of the main features of the micro principle of the social autonomy of the individual is now in order. To begin with, the micro principle is to be understood here as the projection of the Kantian notion of autonomy onto the plane of the social. Autonomy for Kant is the "autonomy of the will."[12] It is essentially a moral concept whose domain is limited mainly to the sphere of private life. In Kant's formulation, obedience to the moral (universal) law given to oneself freely is the source from which the individual draws his autonomy. Obeying the moral law frees the individual from becoming an object of an external action (e.g., desires invoked by others or by unforeseen circumstances). It is possible to project Kant's original formulation of the concept of autonomy onto the legal-political sphere and arrive at the notion of "legal-political autonomy." This notion can be interpreted as the individual's freedom from becoming the object of unlawful external actions (e.g., the authority's arbitrary intervention in her life). That is to say, so long as the individual freely accepts and obeys the law, she is autonomous, in that she is left alone by the state and is protected against the unlawful actions of others. In both private and legal-political spheres, autonomy is the state of being in control of one's situation within the confines of the internalized moral or legal-political laws. Heteronomy, on the other hand, is the state of lacking control over one's situation and becoming an object to be acted on or to be controlled by external forces (others). The concept of social autonomy being advanced here follows in the path of the Kantian original formulation in this respect. However, it differs from Kant's in that here "social autonomy" is defined positively. That is to say, social autonomy is not the state of being completely left alone with one's self-legislated moral laws; nor is it the case of being protected against heteronomy via obedience to the legal-political law. Rather, it is the state of being able to exercise influence and decision-making power over one's situation in micro social milieus. Social autonomy is an empowering and positive notion in that it is both attained and maintained through the social practice of the individual citizen. Despite this major difference, social autonomy follows the Kantian formulation in tying autonomy to obligation. That is to say, in order to attain or maintain his social autonomy, the individual is obligated, or rather "compelled," to interact with the social.

Both the macro principle of the political sovereignty of the individual and the micro principle of the social autonomy of the individual will be discussed in greater detail later in this chapter. What is more important at this point is to call attention to two main features of the "conception of democracy as the political empowerment of the citizen." First, this conception sidesteps the questions of property rights and distributive justice—as it leaves the existing relations of production and property rights assignments as they are, and requires no radical reorganization of society in this sphere. Nevertheless, the conception in question works with the implicit assumption that these relations and rights would not interfere with its principles of political sovereignty and social autonomy. No doubt, in the actual world, this would require the limiting of the freedom of the economic society (the market forces) somewhat. The discussion of the question of how this could be done and the extent to which it could be accomplished—as well as the larger questions of economic equality and property rights in general—will be addressed in Chapter 4. Second, closely tied to the first feature, the "conception of democracy as the political empowerment of the citizen" abandons, for strategic reasons, the traditional understanding of democracy as a leveling or equalizing concept. As *The Betrayal of an Ideal* has argued, it was this understanding that had turned the idea of democracy into what Macpherson has characterized as a "class affair."[13] This feature of the "conception of democracy as the political empowerment of the citizen" will also be discussed in detail in Chapter 4.

Moral Foundations of the Conception of Democracy as the Political Empowerment of the Citizen

The fundamental (justificatory) moral principle underlying the conceptualization of "democracy as the political empowerment of the citizen" is the intuitively plausible claim that a society founded on the basis of this conception of democracy would be the most suitable social-political milieu for the realization of the ultimate value of the human universe, and society's *raison d'être*; viz., the unencumbered, free, all around, and fullest feasible positive development of all individuals in the society.[14] Clearly, this claim is tied at its very core to a conception of human essence—(as well as to a conception of human existence, as will be seen later in this chapter)—which takes Man, using Macpherson's characterizations, as a "developer," and "enjoyer," and an "exerter" of his human capacities, as well as a maximizer of his powers.[15] Moreover, one should note that this contention takes an instrumentalist approach to the question of justifying democracy by treating democracy as the social-political means (institutional designs and social policies) for the end of human development.[16] In this respect, the "conception of democracy as the political empowerment of the citizen" differs from those conceptions that regard democracy, first and foremost, as an "end in itself," or treasure it primarily for what they regard as its intrinsic values (civic education and the social-psychological development of the individual, communal bonding, and benefits to the community).[17] Notwithstanding its primarily instrumentalist view of the question, one needs to add that the "conception of democracy as the political empowerment of the citizen," as will be seen shortly, also regards the practice of democracy as having intrinsic values in rendering direct and immediate benefits to the community. These aspects of the question of justifying the conception of democracy as the political empowerment of the citizen will be returned to later in this chapter. What is more relevant to take up at this point is the exploration of the relations between the political value of direct democracy as defined, on the one hand, and the traditional political values of (negative) liberty, equality, and community, on the other hand.

It is a truism that political philosophers have traditionally privileged the political values of (negative) liberty, equality, and community as primary values and have posited them at the foundations of their theories, as well as regarding them as the primary goals that ought to be realized by society.[18] Broadly speaking, liberal theorists of various persuasions (classical, contemporary classical, egalitarian) have argued for their views and theories primarily from the standpoint of (negative) liberties and the individuals' rights to them. Arguing from concerns for equality and community has also had its place in the liberal tradition, but only insofar as such argumentation strengthened the liberals' positions and deflected the criticisms of their opponents. Consequently, equality and community have played subordinate roles in this tradition.[19] Egalitarians, e.g., socialists and social democrats, on the other hand, and again broadly speaking, have traditionally put equality as the prime value and have appealed to liberty and community as complementary values. Finally, those arguing from the perspective of the common good, e.g., modern day communitarians, have put community before liberty and equality, and have allowed consideration of the latter values, but only to the extent that they strengthen the common good, or do not stymie its realization.

The relevance of this note to the "conception of democracy as the political empowerment of the citizen" lies in that the macro principle of political sovereignty and the micro principle of social autonomy take the trio of liberty, equality, and community as historical facts and as empirically present, and at the same time, as their own theoretical prerequisites. In regard to taking the trio as pre-given, it should be recalled that this conception of democracy was offered against the backdrop of the present-day American liberal-democracy.[20] It is a known fact that the trio in American liberal-democracy comes in a complex package that assigns a higher importance and value to (negative) liberties than to (substantive) equalities, and at the same time regards (formal legal and political) equalities as having higher importance and value than the community.

As to regarding the trio as the theoretical prerequisites for the macro principle of political sovereignty and the micro principle of social sovereignty, it suffices to argue that the exercise of sovereignty and social autonomy would not be possible without first having the essential components of these values actualized in the social-political and legal institutions of society.[21] Alternately stated, the trio should be regarded as the fundamental and the building-block values for the realization of the value of democracy conceptualized by the macro principle of political sovereignty and the micro principle of social autonomy. To support this claim, one can argue that, despite their fundamentality, the values of liberty, equality, and community, without democracy, do not, and cannot, sustain, by themselves, a system of social organization that is primarily devoted to the goal of the fullest feasible and positive development of the human individual. To develop herself fully, the individual needs to be empowered, and as will be argued below, the components of the trio are not in themselves instruments of power, nor can they be directly transformed into such instruments. Rather, they should be seen as enabling and sustaining elements for political empowerment.

This contention will be elaborated further starting with the following paragraph. For the time being, one should argue that the state of democracy in present-day American society lends factual support to this argument. In this society, only the first value in the trio has been realized somewhat maximally; whereas, the second (equality) only partially, and mainly in the political-legal spheres, and the third (community) only minimally, and mainly in the form of patriotism. Finally, one can argue that the fuller realization of the latter two values within the existing liberal-democratic framework would offer to the individual only a more equitable *access* to the agency of power, i.e., the government, and would not necessarily empower him to exercise direct power over this body.

Now returning to the discussion of the moral justifications for the "conception of democracy as the political empowerment of the citizen" that was presented in the beginning of this section, one can argue that the exercise of direct sovereignty or social autonomy by the individual (citizen), in addition to being an instrumentalist exercise in nature, is also an exercise in self-development. That is to say, in addition to serving as the means in the hands of the individual in her attempts to realize her life plans (both socially-regarding and self-regarding ones), acting as sovereign or acting autonomously is directly beneficial to the individual's social-psychological development and citizenship education. Through her acts of exercising sovereignty and social autonomy, the individual comes to learn about her community writ large, and her immediate social milieu as well. She also comes to learn how they function, and hence, develops her views on how they can be improved. In the process, she also bonds with her community at both levels. Furthermore, the exercise of sovereignty and autonomy, as defined, is an exercise in world-making for her—both at the macro levels and in the immediate micro venues. When she acts directly as the sovereign or acts autonomously, she makes things happen; she asserts herself in the world; she shapes the world at large and the one immediately around her. She expresses herself. And she learns that she counts as somebody, somebody powerful and important. She also learns that, far from being a passive subject of the laws that she abides by, she is their co-legislator. It is in this sense that the acts of exercising direct sovereignty and social autonomy turn out to be also directly self-developmental acts insofar as the social-psychological development and citizenship education of the individual is concerned.

There is another aspect to citizenship education that the individual receives in the process of acting as a sovereign or acting autonomously. The exercise of sovereignty or social autonomy allows the individual, the self-developer, the exerter, and the maximizer, to shape the world in ways that would make it conducive to achieving his developmental goals and realizing his potentials and capacities. In order to act directly as a sovereign and act autonomously, the individual needs to be endowed with a certain package of liberties, rights, and a set of equality principles to assure him that he is not hindered in his attempts to achieve his goals and that he is not put into any social-political disadvantage in relation to others. These liberties, rights, and equalities do not empower him (for they are negative in character); rather, they only offer him a framework within which he can seek empowerment by acting as a sovereign and acting autonomously. Moreover, by acting as sovereign and autonomously, the individual learns that these liberties and equalities are designed so that the good of the community is not sacrificed to his, or to others', private wants and desires. He also learns that he will not be able to realize a good portion of his private goals and self-developmental projects to the fullest possible extent if the well-being of the community is compromised. Moreover, he learns that despite his desires to be independent, he is interdependent and he needs the cooperation of others to realize most of his self-regarding and self-developmental plans and goals. Furthermore, he, to use J. S. Mill's words, "learns to feel for and with his fellow citizens and becomes consciously a member of a great community."[22] Finally, he learns "to feel himself one of the public, and whatever is for their benefit to be for his benefit."[23] This *understanding* also leads him to partake in projects and plans that are essentially socially-regarding.

In viewing the exercise of direct sovereignty or social autonomy as a self-developmental and educational exercise in itself, one cannot but tie the "conception of democracy as the political empowerment of the citizen" to a larger tradition in the democratic theory that locates the value of democracy primarily in the direct and immediate educational benefits it delivers to the individual and the immediate community. To those who interpret Rousseau's conception of participatory democracy in a positive light, Rousseau is a champion of this tradition.[24] Among other prominent figures in this tradition, one should mention J. S. Mill, John Dewey, and Alexis Tocqueville. In the case of Mill, his declaration that "the most important virtue any form of government can possess is to promote the virtue and intelligence of the people themselves" is widely quoted.[25] This passage rightly emphasizes the importance Mill assigned to the educational "utility" of democracy.[26] John Dewey was also keen on emphasizing the educational values of democracy, as well as emphasizing the mutually energizing relations between democracy, on the one hand, and education and science, on the other hand.[27] Hilary Putnam's claim that Dewey's conception of democracy can be viewed as an "epistemological justification" for democracy captures Dewey's view on the educational values and utility of democracy.[28] In Tocqueville's case, some of the passages in his *Democracy in America* reveal how closely his views on the relations between the political education of citizens and their participation in political activities are aligned with Rousseau's.[29]

Finally, before closing this section, one needs to add three relevant comments on the question of providing moral justifications for the "conception of democracy as the political empowerment of the citizen." First, it should be noted that the instrumental justification offered earlier in this section goes well beyond what is often referred to as the "protective" justification for democracy. Protective justification is often offered as the main defense for liberal democracy. It locates the value of democracy mainly in protecting the civil-liberties and rights of the individuals and groups.[30] Second, it should also be commented that, in addition to the instrumentalist moral justification offered above for the "conception of democracy as the political empowerment of the citizen," there exists another moral justification for this conception. That is to say, one can argue that the conception in question also posits democracy as encompassing two other values. The first is the value associated with the idea of respecting each and every individual as a worthy member of society, and at the same time, as an equal in status and power to every other member. The value in question is affirmed in the idea of empowering the individual to participate directly and actively in political decision-making and legislating. This is in essence a Kantian justification. The other value encompassed by the "conception of democracy as the political empowerment of the citizen" is the value of the self-government or autonomy of the community (both in micro and macro communities). This value is affirmed in treating the community as the source of laws and social policies.[31] (This contention will be returned to later in this chapter.) Lastly, one should comment that it is possible to construct a Rawlsian sort of contractarian argument in support of the "conception of democracy as the political empowerment of the citizen"; that is to say, to argue that the participants in the Rawlsian original position would agree to this conception of democracy alongside the two principles of justice they adopt.[32]

A Democratic Conception of the Individual

The conceptualization of the idea of democracy as the political empowerment of the citizen is in a sense the conceptualization of the idea as an *individual affair*. This approach to the question differs considerably from the approaches that take democracy as a "class affair" (e.g., the approach of Macpherson, Marx, and Lenin), and from Rousseau's which takes democracy as an *all-people affair*.[33] The conceptualization of democracy here as an individual affair follows in the tracks of the "methodological atomic individualism."[34] Despite embracing atomic individualism in methodology, the "conception of democracy as the political empowerment of the citizen" cannot but reject the types of moral or political individualism that are often attributed to liberalism. Broadly speaking, what lies at the root of liberal individualism is a conception of the individual that posits her as primarily, if not purely, a self-interested, self-regarding, egoistic, and isolated being.[35] Posited as such, the liberal conception of the individual is incapable of meeting the prerequisites of political or social empowerment as defined earlier. To begin with, the macro and micro principles proposed earlier would require a conception of the individual that takes her as being endowed with intrinsic socially oriented attributes in addition to the self-oriented ones. In such a conception, the individual would be posited as self-interested yet interdependent, egoistic yet intrinsically socially-responsible (and not necessarily altruistic). Moreover, the macro and micro principles would require that the individual be conceptualized as being conscious of her self-interests, and at the same time, as being a socially-conscious chooser.

Furthermore, the conception of the individual that could be theoretically compatible with the macro and micro principles would regard the individual's political participation and choice-making in public affairs as being motivated by her self-interests, yet being informed and conditioned by her social consciousness and the *understanding* that her personal interests can be best pursued and attained in a society committed to her well-being, as well as being committed to the well-being of all. Without a well-ordered and well-maintained society, and without the cooperation of other individuals, she *understands* that the realization of her self-oriented pursuits will be frustrated or stymied. Thus, she recognizes that her relations with other individuals are as important to her well-being as her own self-interested plans are to her. In short, the conception of democracy being developed here would require a conception of the individual that posits her not merely as a self-interested and self-oriented being with a set of prefigured and fixed preferences and interests; but also as an inherently social, socially oriented, and socially conscious being. That is to say, a social being who *understands* the *interdependence* of, and the *interconnections* between, her private interests, on the one hand, and the common interests of the community, on the other hand.[36]

As one can see, the conception of the individual being sought here would necessarily be a synthesis that draws from both the liberal tradition, and those conceptions that regard the individual as essentially, or exclusively, a social-being. It is a truism that conceptions in this latter category tend to minimize or deny what Carol Gould calls the "ontological primacy" of the individual through subjugating her to community. This subjugation takes place in numerous ways—theoretically, politically, morally, and rationally—and in varying shades and degrees. Among these conceptions, one should mention Communitarianism and most Marxist trends, as well as some trends in the theory of deliberative democracy. Some trends in Marxism do this subjugation theoretically, others politically.[37] Communitarians do it morally and politically. One cannot emphasize enough that what makes the synthesis of the concept of the individual being attempted here possible is the *integration of consciousness as a component into the conception of the individual.*[38] This synthesis postulates the human individual as a *fluid* and complex unity of self-orientedness, on the one hand, and sociality, on the other hand. She is *both* an isolated individual, and a social-being; an ensemble of individualistic and selfish interests, and at the same time, an "ensemble of the social relations"; *both* potentially a slave to her (natural) selfish instincts, desires, interests, passions, and an agent of instrumental rationality; and at the same time, a potential social martyr and a potential Kantian rational agent—i.e., reason and morality could constitute or rule the content of her wills, ends, or her (personal or social) choices. Her actual and concrete being is a compromise and an on-going negotiation between these two aspects that constitute her as a being in the social world.[39] She is a being capable of both *understanding* the importance of her sociality, and of being persuaded rationally and morally, as well as pragmatically (instrumental rationality). In one respect, she is the Rawlsian "reasonable" person with the added attribute that she has a much higher level of social consciousness than Rawls is assuming.[40] In short, she is not a purely abstract being constructed in a philosophical plane on some benevolent notion of human nature, nor is she the purely self-oriented, "possessive," and greed-driven individual of classical liberalism or the Hobbesian state of nature; but rather a complex being who has attributes from both worlds—and who is *conscious* of this duality that constitutes her being. As posited, she, in some respects, is the actually-existing individual of present-day American liberal democracy who has a good grasp of what Tocqueville termed the "doctrine of self-interest well understood."[41] The complexity of her existence is a reflection of the complexity of her "nature"; and, at the same time, the complexity of her nature is a reflection of the complexity of her existence. This synthesized conception of the individual is of utmost importance for the "conception of democracy as the political empowerment of the citizen" being formulated here. In addition to assigning to the individual's empowerment a social character, and thus providing a basis for negotiation and compromise among the self-interested individuals, the synthesized conception of the individual also offers a basis for mustering and promoting social cooperation among individuals as socially oriented and socially conscious citizens.

The contrast between positing the individual as self-oriented versus positing him as socially oriented has traditionally resonated and paralleled the contrast between the notion of actual (selfish) Man versus that of a benevolent real (or potential) Man. In Macpherson's modeling of this contrast, the former notion is characterized as "descriptive"—(in which Man is regarded as selfish, possessive, and a personal utility maximizer)—and the latter as "ethical"—(in which Man is posited as a maximizer of his essential powers, "developer," "exerter," and "enjoyer" of his "human capacities").[42] This contrast has also been framed within the context of opposition between existentialist conceptions of Man (e.g., Jean Paul Sartre's) and the essentialist conceptions (regardless of whether Man is regarded as a benevolent or non-benevolent being). Needless to say, the synthesis in the conception of the individual reached above would necessarily need to be supplemented with an analogous synthesis in the conception of human "nature."

To this end, one can begin by arguing that the synthesis in the conception of the individual offered above amounts to an essentialist conception of Man that takes him as having an essence and at the same time, it insists that this essence is a fluid one. This fluidity leaves the essence open to the influences in his milieu of existence. In this sense, human essence is to be understood as being socially conditioned and historically developed. What is truly essentialist about this conception is that Man is assumed to possess certain attributes that define what he essentially is. This essence was defined in the preceding section in terms of the attributes of "developer," "enjoyer," and "exerter" of his "human capacities" and "powers" (Macpherson's terms).[43] Having stated this, one should be quick to point out that these essential attributes and powers, as Macpherson himself believed, are socially constituted—(as well as being historically developed)—and also that the degree and quality of the individual's development, enjoyment, and exertion of these capacities or powers, are directly tied to the existence of favorable social milieus, which in turn, are conditioned and shaped in part by the individual's own mode of existence—notably his political participation and social consciousness.[44] Moreover, as can be recalled, the conception of the individual synthesized earlier, incorporated attributes from both the "ethical" and "descriptive" conception of human nature. Such a synthesis, as Macpherson has argued, is not at all "logically contradictory" or "logically incompatible."[45]

To sum up, in light of the emphasis placed on social consciousness as the key element in developing the conception of the individual attempted above, and given that social consciousness is a socially developed and socially sustained product, the synthesis reached here necessarily took as its point of departure the socially constituted and essentialist side of the contrasting philosophical positions on the individual, and went on to incorporate into itself almost entirely all of the attributes of the individual often found in liberal conceptions. Moreover, it regarded what it took as the human essence as being historically and socially conditioned by the context of Man's existence.

Finally, before leaving this section, given the emphasis placed on social consciousness, the compromise conception of the individual reached in this section would need to be tied to a conception of society that would be conducive to the development and sustenance of both this consciousness and the personality traits that would harbor and nurture it. These traits constitute what Carol Gould calls "democratic personality." She lists the following as the main traits of this personality: "initiative," "disposition to reciprocity," "tolerance," "flexibility and open-mindedness," "commitment and responsibility," "supportiveness, sharing, [and] communicativeness."[46] The following section will present a discussion concerning the kind of society that would be conducive to harboring and spawning these traits as well as possessing some other elements needed for theoretically supplementing the "conception of democracy as the political empowerment of the citizen."

A Democratic Conception of Society

In order for the individual to be able to exercise his political sovereignty and social autonomy, i.e., to be truly a citizen, he needs to be provided with a framework or a certain package of basic freedoms, and the rights to them, as well as with a certain system of social-political equalities. Moreover, the individual would need to be provided with what is needed to develop and maintain the *consciousness* that not only a great portion of his private interests, on the one hand, and public interests, on the other hand, are interconnected, but also that the public interests, in many respects, are part and parcel of his own broad and long-term private interests.[47]

This would require a strong system of liberal and civic education, as well as a culture that cultivates, nurtures, and sustains this consciousness. It should be noted in passing here that throughout this book, the terms liberty and freedom are taken as being equivalent and are used interchangeably. These terms are taken to mean the absence of both physical coercion and "domination by withholding the means of life or the means of labor" as Macpherson has commented.[48] This would require that in addition to being free, the individual must also be made to *feel* free. That is, in addition to being entitled to a system of liberties that would enable him to exercise his sovereignty and social autonomy, the individual would also need to be free from the fear of persecution, reprisal, or the threat of having his means of life withheld from him—which could be evoked by the state or by other individuals. As was noted earlier, it is the contention of this book that present-day American society has established this framework of liberties with some reasonable levels of success.

As to the need for a system of equalities, one can argue that, in order to be capable and motivated to exercise his sovereignty and social autonomy, the individual would need to be on a relatively equal footing with other members of society, both socially (i.e., the principle of equal respect for the opinion of each and every citizen regardless of his social and economic status in the society) and politically (i.e., the principle of giving equal weight to the vote of each and every citizen regardless of his social and economic standing in the society). This amounts to the contention offered earlier in this chapter that the "economic society" ought to be prevented from interfering with the macro and micro principles. The basic idea behind the need for a principle of social equality arises from the need to furnish a social-psychological framework to motivate the individual to be willing to exercise his social autonomy. To do so, the individual would need to be (and feel) as socially significant and socially worthy as the next person. This is due to the fact that at micro levels or local settings, most interactions will be face-to-face. No doubt, the extent of the individual's involvement in face-to-face interactions may vary from one individual to another, and from one social/economic institution or setting to another, depending on the nature and function of the entity and its circumstance. However, the extent of the individual's participation in the exercise of social autonomy ought not, in any form or degree, compromise or threaten the civil liberties of the other individuals involved in that entity. That is to say, the participation ought to be entirely voluntary, and without an actual or consciously evoked fear of reprisal or loss for those who choose not to exercise their social autonomy. No doubt, those who choose not to participate will have to abide by the decisions made by those who do participate. One can argue that this notion of social equality or equality of social status is fundamentally the same as Rawls' idea of "the bases of self-respect" (one of the three social "primary goods").[49] The notion of the equality of social status being put forth here also finds some resonance in Gould's notions of "reciprocity" and "equality in interpersonal relations."[50]

In contrast to the exercise of social autonomy that will rely heavily on a strong principle of social equality, the exercise of political sovereignty will need a strong and substantive principle of political equality (i.e., the vote of all ought to carry the same weight regardless of the social and economic status of the individuals involved). This is mainly due to the fact that at the macro levels, face-to-face interactions will be minimal and the exercise of sovereignty will necessarily be done by voting. In order to prevent the social and economic organizations, associations, and institutions from influencing the political process disproportionately, as now is the case in present-day American liberal-democracy, the domain of the applicability of the principle of political equality should be extended beyond the persons. As to the system of political and social equalities, one can argue that a slightly modified version of Rawls' principles of justice can satisfy this need maximally. Raised on the fertile grounds of the fundamental "intuitive ideas" of "society as a fair system of cooperation" and "citizens as free and equal persons," Rawls' theory of justice has all the fundamental elements necessary for meeting the needs of the reformulation being attempted here. A detailed elaboration of this point extends beyond the scope of this book. However, in passing, it should be noted that the need for the modification of Rawls' theory arises mainly from the necessity of reconciling his theory with the post-liberal conception of the individual offered in the preceding section.[51]

Finally, in order to be able to exercise his social autonomy and political sovereignty, the individual would need to live in a "democratic culture"—Habermas characterizes this culture as an "accommodating political culture."[52] (Without a doubt, the exercise of sovereignty and social autonomy itself would contribute to the further development and strengthening of this culture.) This political culture can be characterized by the prevalence of a high level of social-political literacy and general education, and the currency of participatory attitudes. Furthermore, this political culture would need to be able to inculcate the traits of "democratic personality" in the individual.[53] The political culture in question would need a strong system of civic and liberal education for its sustenance. Moreover, this culture would encompass a broad, yet shared, spectrum of somewhat compatible public beliefs about what constitutes human rights and liberties. Finally, this spectrum would also encompass a set of elastic and publicly respected conceptions of the "good life" and "good society," as well as a set of broad conceptions of values and ends generally regarded as "worthy."[54]

The full investigation of the question of how such a spectrum could be formed, as one can imagine, is a formidable theoretical undertaking in itself and extends beyond the scope of this book. Nonetheless, in passing one can speculate that the process of the development of this spectrum could be guided by the following three considerations. First, in accord with the conception of the individual offered earlier, the "good" is to be understood not so much in terms of the individual's rights to personal preferences or self-orientedness. Rather, it should be taken more in the sense of the "quality" of the goals he pursues as a way of fully developing his essential powers. Clearly, the conception of the "right" implied by this conception of the "good" would take rights as a framework within which the individual will be empowered to pursue his self-developmental goals, which in essential ways would be similar to those of other individuals. Second, consistent with the fundamental assumptions of the conception of "democracy as the political empowerment of the citizen," the scope of what constitutes a right and how far it can be extended should mirror the socio-economic development and material-technological achievement of society.[55] Third, the scope and extent of the rights, and liberties, as well as the values underlying the differing conceptions of the "good life" and "good society" should be—(to use Rawls' phraseology in his justificatory argument for his theory of justice)—"rooted in the basic intuitive ideas found in the public culture of a constitutional democracy."[56] The question of how such a "public culture" could flourish or be sustained will be revisited in Chapter 4 and discussed in some detail.

The Individual Acting as the Sovereign

In the capacity of the citizen, the individual is a public person with public interests, as well as a holder of a set of views and conceptions on the matters of the "common good." He acquires some of these conceptions as part of the general and formal education he receives earlier in life (including civic education and civic skills). He develops these conceptions further and acquires new ones as part and parcel of his general social consciousness. The latter development takes place as the individual interacts with other citizens in the immediate community of his everyday life, and as he communicates with the larger public through the media of mass communication. In these interactions and communications, the individual learns about the issues and matters of concern in the immediate community, as well as in the larger public. In this process, he also develops his own views on what constitutes the good of the immediate community and the larger public in these issues or matters, and how the issues ought to be addressed.

Now, assuming that the individual is committed, and willing, to "act" on his views on these matters, and assuming that these views of his are socially and politically "educated" and "reasoned," then one can argue that his views should be treated not just as his private opinions, but rather as his "*public wills*" on these issues or matters.[57] Moreover, granting these assumptions, one can further argue that the individual would be willing to participate actively in collective decision-making activities on the issues and matters facing the public, provided that there exists a system or scheme of collective decision-making that he, and the public in general, regard as being *"adequately democratic."*

Given this scenario, and keeping in mind the ideal macro and micro principles of democracy proposed earlier in this chapter, one can now define or envision an "adequately democratic" decision-making scheme as the one which empowers the individual to bring her "public wills" to the attention of the public (e.g., in public forums and public deliberations) *and* further empowers her to "conjoin" these wills with the similar wills of her fellow citizens. The objective of this conjoining of the public wills of the individual citizens is to produce what will be referred to from now on as the "*collective wills*" of the community, or the nation, on the issues of public interests. One can further envision that this adequately democratic system or scheme has the complexity or sophistication to accept the expressions of the wills of citizens *directly* and *fully*. Directly, in that there would be no intermediaries such as representatives. Moreover, fully, in the sense that the scheme would first allow the individual citizens to express their wills as they wish, and then would process the expressed wills in the form they are expressed. That is to say, the scheme would process the expressed wills of the citizens with the same degree of "intensity" and the same level of "complexity" with which they were expressed or entered into the scheme. Stated alternately, the envisioned scheme or system would provide each citizen with the power of "*full expression*" of her wills *and* the power of "*full integration*" of these expressions into the collective decision-making process. Finally, it is assumed here that in "conjoining" the expressed wills of citizens in order to "amalgamate-compose" a collective will on each decision-making occasion, the envisioned scheme would use a complex procedure that would be *broadly regarded by the public as being a "reasonably fair and accurate" method* for amalgamating-composing a collective will (or a public decision) out of the wills entered into it by individual citizens. In the rest of this chapter, this envisioned scheme will be referred to as the *"idealized"* scheme of collective decision-making.

With these in mind, one can advance two arguments, both of which are intended to tie the idea of "individuated sovereignty," discussed earlier in this chapter, to the "idealized" scheme of collective decision-making envisioned above. These arguments will be presented below and will be immediately followed by a discussion of some of the assumptions that they implicitly entail.

The first argument is that the individual's participation in the "idealized" scheme of collective decision-making can be said to constitute the direct exertion of her "share" of the political sovereignty. In other words, on each public decision-making occasion, by expressing her conception of what the collective will or public policy ought to be, *and* by directly entering this conception into the "idealized" collective decision-making scheme—that would eventually produce or compose a "collective will" on the issue in question—the individual directly exerts her share of "individuated sovereignty." That is to say, the individual exerts her portion or share of the sovereignty of the people that has been divided equally among politically and legally free and equal citizens.[58]

Now the second argument: one can further argue, given that each and every citizen is equally endowed with the same set of rights and powers in the "idealized" collective decision-making scheme, the resulting collective will, on each decision-making occasion, would be the embodiment of the sovereignty of all (i.e., the sovereignty of the whole community or the sovereignty of the people), and thus would constitute a true expression of the rule by the people. Indeed the collective wills or decisions produced by the "idealized" scheme could truly and legitimately lay claim to being *democratically formed wills* or *democratically decided policies or laws*.[59] What follows is a discussion of some aspects of the "idealized" scheme which is intended to bolster the second argument.

To begin with, the condition in the "idealized" scheme, that the individual brings his public wills to the attention of the public, is intended to facilitate the social-political education of his wills before they are entered into the decision-making scheme. By expressing his conception of what the collective will ought to be on each major issue before the public, the individual citizen hopes to persuade his fellow citizens to see the merits of his conception; and at the same time, he leaves himself open to the possibility that his own conception could be altered or revised as a consequence of being persuaded by the arguments of his fellow citizens. This would require the existence of a comprehensive set of public deliberation and civic education institutions. Clearly, the existence and proper functioning of these institutions would enhance the legitimacy and public-acceptability of the collective wills produced by the scheme. That is to say, on the strengths of these institutions, the collective wills produced in the "idealized" scheme would lay claim both to representing the general reason and judgement of the public, and the pooling of its collective wisdom, as well as to being the embodiment of the educated and reasoned will of the public.

Furthermore, by giving the individual citizen the power of the *"full expression"* of his wills *and* the power of the *"full integration"* of these expressions into the collective will-formation scheme, the collective decision-making scheme fully "hears" the wills of the individual with all of their complexities, and fully considers and "integrates" them into the amalgamating-composing procedure that would result in producing collective wills. This capacity of the "idealized" scheme fully empowers the individual to exert his share of sovereignty to the fullest possible extent. The importance of this consideration is that it further enhances the legitimacy of the collective wills or decisions produced by the scheme. The individual would accept the legitimacy of the formed collective wills (even if the resulting "collective will" turns out to be the opposite of what he brought into the process as his input), and thus would find himself morally persuaded to abide by them, for his own wills or inputs into the process of producing the collective wills were fully considered and accounted for in the scheme that produced them.[60] Of course, the ultimate criterion of the legitimacy of the collective wills or decisions produced by the scheme lies in the assumption that the scheme would use, as noted earlier, an amalgamating-composing procedure that would be broadly regarded as being a "reasonably fair and accurate" method for forming or composing the collective wills out of the entered wills of the individuals. A full discussion of how such a procedure could be devised will be presented in Chapter 2.

To recap, the focus of this section so far has been on tying the idea of the "individuated sovereignty," discussed earlier in this chapter, to the "idealized" scheme of collective decision-making that was introduced in this section. This scheme provides citizens with a set of public institutions for expressing their public wills and discussing them—(mainly for the purpose of educating their wills)—and then enables them to enter these wills as they wish—(with the complexity or the degree of intensity they deem appropriate)—into a procedure that considers them fully in a "reasonably fair and accurate" process of amalgamating-composing collective wills out of them. Contingent upon the general acceptability of this procedure as being "reasonably fair and accurate" by the public, an attempt was made to argue that the collective wills produced by the scheme would have democratic legitimacy. The "idealized" scheme was envisioned as a way of giving a plausible expression to the idea of the direct exertion of the "individuated sovereignty." One needs to emphasize here that notions of the "full expression" of the wills of the individual and their "full integration" into the collective will-formation scheme are the most fundamental components of the "conception of democracy as the political empowerment of the citizen" being advanced here. One also needs to add that these notions are also intended as optimal notions and not as absolute ones; that is to say, as notions that are conditioned by feasibility and the technologically available means of communication and schemes of social decision-making. It is the contention of this book that such technological means are available in present-day society. A discussion of these means will take place in Chapter 2 and Chapter 3.

* * * * *

As a way of strengthening the appeal of the "idealized" scheme of the collective decision-making envisioned above, a further elaboration of what is meant by "wills" in this chapter is in order here. To begin with, a "collective will" is a collectively decided will that emerges in the process of conjoining the wills that the citizens enter directly and individually into the process in an attempt to collectively form a public will on a given public decision-making occasion. As to the wills of the individual, the "public wills" are taken to be the expressions of the individual's conceptions of the common good and her political-moral outlooks on the issues facing the society, as well as being the expressions of her criteria of rationality and truth. The individual's public wills are also taken to be the applications of these conceptions, outlooks, and criteria to the issues at hand. Moreover, public wills are what the individual wills to be actualized in laws or in public policies. "Public wills" differ from mere views and opinions of the individual, in that she is willing to bring them to the attention of the public and suggest them as possible courses of action to her fellow citizens. Finally, the public wills of the individual differ from her mere opinions in that the individual is aware that her public wills would impact the lives of her fellow citizens, as well as her own (if materialized in laws or policies); and thus she understands that her public wills carry with them the weight of her social responsibilities.

The intent of the individual's "public wills" is the good of the larger public and her immediate community. They are the expressions of both her willingness to participate in the affairs of the community and the nation, and her (politically and morally based) preferences for the courses of action that the nation and the local community ought to take, and the choices they ought to make regarding the issues to be decided publicly. Given that public wills aim at the good of the public, they are fluid and open to the possibility of being altered or transformed by the morally or rationally grounded arguments that have the realization of the common good as their objective. Finally, considering that these wills are in a constant state of being formed and reformed—due to the fact that they are publicly debated—the individual's public wills ought to be regarded as being the *expressions of considered and reflected-upon judgements* of the individual on public issues, and at the same time, as having more consistency, substance, and force than mere personal opinions, preferences, choices, or tastes—especially in light of the fact that the individual is willing to "act" upon them. Preferences, unlike wills, are often fixed, rigid, pre-given or pre-determined, desire-based, and un-examined (as often is the case with one's cultural espousal of her heritage, such as her taste in food), and thus cannot serve as the basis of political decision-making.[61]

The individual's "private wills," on the other hand, are the expressions of his preferences and choices on the same issues, but this time, expressed from the perspective of his own private interests. As defined, the individual's public wills, no doubt, are distinct from his private ones. His private wills may or may not be in harmony with his public wills. Moreover, he may, or may not, be conscious of the inter-connections between his private wills, on the one hand, and his public wills, on the other hand. It is a well-known fact that in expressing their wills in elections, the individuals in American liberal democracy do not and—given that they express their wills monosyllabically (by saying yes or no, provided that they do get the opportunity to express their wills)—cannot separate their private wills from public ones. The benevolent assumption in the United States that the individual casts, or ought to cast, his vote purely on the considerations of the public good, in light of this difficulty, is a naïve and utterly farcical assumption. It is common knowledge that voters in this society let their private wills overshadow and influence their public ones.

Now, given that the "conception of democracy as the political empowerment of the citizen" is erected on the notions of "full expression" and "full integration" of the individuals' wills, and given that the individual possesses both types of wills, one is led to argue that the private wills of the individual would also be considered, to some extent, in the process of the formation of the "collective wills." Without a doubt, arguing for this point amounts to making a radical break with Rousseau and the theories of participatory and deliberative democracy—all of which are premised on transforming, superseding, and sublimating the private wills of the individual. (This is often done by overburdening the individual, either rationally or morally.) Thus, within the conceptual scheme of "democracy as the political empowerment of the citizen," the "collective will" formed in each public decision-making occasion turns out to be a composition that reflects the sum total of all wills—i.e., both the public and private wills of all citizens. Thus understood, *the notion of "collective will" here is different from Rousseau's "will of all"* (which is "a summation of the private wills") *and "general will"* (which is the resultant of only the public wills).[62] The answer to the question of how the public wills could be combined in a consistent manner with private wills was hinted at in the course of formulating the democratic conception of the individual earlier in this chapter. That is to say, social consciousness and an "enlightened" conception of private interests similar to Tocqueville's "doctrine of self-interest well understood" could make it possible to find some consistent and meaningful ways of combining the citizens' private wills with their public wills.[63] (The question of the extent to which private wills could be considered in the formation of the collective wills will be discussed in Chapter 2.)

As a way of completing the formulation of the conception of the "individuated sovereignty," one needs to discuss how this conception differs from the now-ubiquitous notion of "consumer sovereignty." To begin with, while the latter is essentially an economic concept that has gained currency within the domain of the market-based conceptions of society, the former is a political-moral notion. Individuated political sovereignty has as its main object both the political empowerment of the individual (as a citizen) and the good of the community. That is to say, it takes the individual as a public person, i.e., as a citizen, or as a *political-choice maker* who has public wills, concerns, and interests, on the one hand, and public obligations and responsibilities, on the other hand. In the notion of the consumer sovereignty, in contrast, the individual is posited as a private person, i.e., a *"rational-choice" maker* with private wills whose main concerns and interests are self-regarding, and economically based. The object of consumer sovereignty is to maximize the command, and extend the sway of, the individual's *self-regarding preferences* and financial resources in the marketplace. By contrast, what the individuated political sovereignty aims at is the promotion of the *public wills* and aspirations of the individual on matters or concerns that he, as a *public person* or a *citizen*, shares with other citizens.

The distinction just drawn between the individuated political sovereignty, on the one hand, and the consumer sovereignty, on the other hand, is not purely theoretical. There exists ample empirical evidence to support the claim that this distinction is actually present in the public consciousness and political behavior of the citizens of present-day society. Cass Sunstein's comments in his recent book, *Republic.Com*, on the actual behavior of the individuals in this society depict the situation eloquently. The whole passage is worth quoting at length.

> Citizens do not think and act as consumers. Indeed, most citizens have no difficulty in distinguishing between the two . . . [T]they seek stringent laws protecting the environment or endangered species, even though they do not use the public parks or derive material benefits from protection of such species; they approve of laws calling for social security and welfare even though they do not save or give to the poor; they support antidiscrimination laws even though their own behavior is hardly race- or gender-neutral. *The choices people make as political participants seem systematically different from those they make as consumers.*
>
> . . . people's behavior as citizens reflect a variety of distinctive influences. In their role as citizens, people might seek to implement their highest aspirations in political behavior when they do not do so in private consumption . . . people might "precommit" themselves, in democratic processes, to a course of action that they consider to be in the general interest. And in their capacity as citizens, they might attempt to satisfy altruistic or other-regarding desires, which diverge from the self-interested preferences often characteristic of the behavior of consumers in markets. In fact social and cultural norms often incline people to express aspiration or altruistic goals more often in political behavior than in markets. Of course selfish behavior is common in politics; but such norms often press people, in their capacity as citizens, in the direction of a concern for others or for the public interests.[64]

Another aspect of the difference between consumer sovereignty and the individuated sovereignty is that in the former, sovereignty is understood in the sense of the consumer directing the producer to produce her desired items of consumption and do so in accordance with her preferences. Viewed from the other end of the relation, the producer takes orders from the consumer, and does so gladly, for doing so would maximize his gain. In the individuated notion of political sovereignty, in contrast, there exist no analogous relations. The individual citizen places no orders. She herself is a *co-producer* of the laws, some of which she would gladly abide by, and some others which she would obey as her obligations as a citizen.[65]

Finally, before closing this section, it would be relevant at this point to call attention to the fact that unlike the "idealized" scheme, the existing collective decision-making arrangements in present-day American democracy fall miserably short of allowing the wills of the individual to be fully expressed or directly integrated. To start with, the individual in this society almost never gets the opportunity to express his wills on the issues, whether fully or partially, except in plebiscites, referenda, initiatives, and recalls, which are extremely rare occurrences (perhaps once in a lifetime of a citizen).[66] The only real opportunity he has in the way of expressing his wills is to surrender (transfer) his will-formation and will-expression powers to some other individuals, i.e., to his virtual-independent representatives. He also has a potential opportunity in this respect, which basically consists of attempting to influence the decision-making patterns of the representatives (either by writing to them, or in rare cases, meeting them in person). As will be discussed later, this severance from experiencing political power *directly* is one of the main sources of *political* alienation in present-day American society.

The Individual Acting Autonomously

The core idea underlying the principle of social autonomy is that of exercising sovereignty within the domains of one's everyday life, i.e., having power to influence these domains; and being in control of one's situation to the highest degree feasible, and without a doubt, to the extent that it is compatible with the same degree of autonomy for the others in the domains in question. Social autonomy is the ability to make one's social activities an object of deliberation within the social settings where the activities take place. It is a principle for equal power-sharing and equal partnership at micro levels. Moreover, social autonomy is the individual's independence in dealing with others in social groups beyond the necessary levels of interdependence present among the members in every group. Social autonomy is an expression of equal regard and dignity for each individual in contributing to decision-making at micro-levels. Thus defined, social autonomy is the opposite of social heteronomy, i.e., the state of lacking control over one's social activities and situation, and thus being an object to be acted upon or being controlled by external forces (e.g., by hierarchical institutions or other individuals).

Within the confines of frameworks that determine or condition activities in the domains of his everyday life, e.g., the workplace, the individual exerts his social autonomy; that is to say, he partakes in the activities that give laws to him and to the domains in question, through the exercise of judgement, choice, and responsibility.

Exertion of one's social autonomy in the small social settings of everyday life, in which one works or socializes, can be modeled very much in the same way the exertion of sovereignty was conceptualized earlier. What is different in these domains is that it is not the larger public interests, but the immediate private interests of the individuals, as well as those of the social group in question, that reign supreme. As a group member with private interests, the individual is faced with issues that figure prominently in his everyday life at the workplace, his immediate neighborhood, and the social associations and settings of which he is a member, either out of necessity (e.g., the school which his children attend), or voluntarily (e.g., church or a neighborhood association to which he belongs). The exertion of social autonomy by the individual in these settings would mean being involved in collective decision-making situations that would rely more on face-to-face interactions and less in casting his vote from a distance, as is the case with exerting his sovereignty at macro levels.

Thus defined, social autonomy is not the same as personal autonomy, i.e., full self-mastery or full self-direction, which is only relevant to the private domains of one's life (of course, within the confines of the necessities that surround one's private life), e.g., the domain of one's ethical life in the Kantian sense. Rather, social autonomy denotes the social power of the individual to contribute to managing and giving directions to the social units wherein he is a member, as well as denoting his power and right to participate in maintaining the well-being of these units. The degree and scope of these powers are negotiated internally among the involved parties in the domain or association, on the one hand, and are bound by the framework of guidelines and structures provided by the appropriate local, state, or federal institutions or governing bodies which the domain or association is connected with, on the other hand. Needless to say, one would assume that this framework, as part and parcel of the larger societal framework, respects and guarantees the rights and benefits granted to each and every individual by the constitution (e.g., civil liberties and equal rights). As one can recall, in the case of the "individuated sovereignty," the scope and degree of the individual's power to legislate, as well as the range of what could be subjected to collective decision-making, were also bound by a framework (the constitution in this case). Moreover, as was the case with the individuated political sovereignty, or the idea of democracy itself, social autonomy is posited as a "matter of degree," and is not a quality that an individual either possesses or lacks.

As an alternative approach to formulating what social-autonomy represents, one can view the idea as a context-specific, locally conditioned, and collectively defined type of self-determination. That is to say, the individual is self-determining in that he has some substantive powers and rights in shaping or influencing the content and form of the activities he participates in at micro levels.

These powers and rights are compatible with the same powers and rights for others involved at the same level. Without a doubt, these powers and rights ought to be defined so that they would not stymie the proper-functioning of the micro units in question. As a worker on the shop floor, or a manager in a factory, or as a staff member in an office setting, the individual would have some substantial powers over, and input into, the processes of performing his responsibilities, as well as having some say and stakes in the overall performance and well-being of the firm or social grouping to which he belongs. Thus, social autonomy is not merely an attribute of the individual, but mainly that of his social unit or association. That is to say, social autonomy is essentially more socialistic or collectivistic than individualistic.[67] Its scope and powers are constrained immediately by the procedures and rules of the association itself, and through those, ultimately by the larger framework and the democratic structure of the whole society to which the social unit belongs.[68]

Social autonomy, as defined, can be conceptualized as miniaturized sovereignty (or empowerment) that can be exercised at the micro levels of one's everyday activities or immediate community. This would mean that the notions of "full expression" and "full integration," as well as the distinctions between the "public wills" and "private wills," would also be relevant to the question of the exertion of social autonomy. Moreover, this would imply that a modified version of the "idealized" scheme could be employed for collective decision-making purposes at these levels. A major difference between social autonomy and the "individuated sovereignty," as was noted earlier, is that social autonomy is primarily concerned with the private interests of individuals in small social units. Another difference between the two is that, given the predominance of the face-to-face interactions at micro levels, deliberative modes of thinking and discourse would acquire greater importance for social autonomy. Thus conceived, micro levels would constitute the main domain of applicability for the theories of deliberative democracy.

However, considering the possibility that at micro levels "orators and philosophers" could easily hijack the deliberation sessions or could manipulate others and talk incessantly, some individuals might choose to minimize their participation in deliberation sessions at the workplace or at local community organizations, or stay away from them altogether. For this reason, and for a variety of other reasons that will become apparent later in the book—(and contrary to what the theories of deliberative democracy advocate)—the micro deliberation sessions should not be conceived or modeled primarily as decision-making sessions. [69] Thus, failure to participate in deliberation sessions, as was the case with the exercise of the "individuated sovereignty" at the macro levels, should not strip the individual of his rights to express his wills and to have them incorporated into decision-making at micro levels. For these reasons, expression of the wills here, as was the case with the exertion of the "individuated sovereignty," ought to be perceived as an activity that would take place in a private space and moment by the individual.

Concluding Remarks

This chapter attempted to formulate a conception of democracy that holds steadfastly to the original meaning of the idea of democracy—i.e., the idea that democracy is about demos being sovereign and exercising this sovereignty directly; that the people ought to be the legislators of the laws they abide by. In the conception formulated above, the proof of the sovereignty of the people does not lie in that their "general will" is sovereign in the state, but that they *themselves*, as individuals, are the sovereign; that they *themselves are the actual decision-makers*. This conception was referred to as the "conception of democracy as the political empowerment of the citizen." The conception in question was formulated in terms of two fundamental principles of the "individuated political sovereignty" (the macro principle) and the "social autonomy" (the micro principle) of the individual (citizen). Moreover, this formulation took place within the political-theoretical framework of present-day American liberal democracy. What is novel about this conception is that its first principle posits sovereignty of the people as the force of the "collective wills" that emerge from an idealized process of amalgamation-composition of the wills expressed by each and every individual (citizen). A collective will that emerges in this process would legitimately have the status of a democratically decided policy or law. Thus conceived, the force of the wills expressed by each individual citizen in this process would be tantamount to the exertion of his individuated sovereignty. Moreover, the very process of the collective will-formation would then become the actual process of exertion of the sovereignty of all (the people).

In the same vein that the first principle empowers the individual to contribute directly to, and exert influence in, the process of making macro level political decisions, the principle of social autonomy enables the individual to exercise influence in, and give direction to, those social-political factors that shape or affect his everyday life at the workplace or at social groups to which the individual belongs. As a way of revealing the inner connections between these two principles, one might argue that, while the individuated political sovereignty can be viewed as social "autonomy writ-large" (i.e., at macro levels) or as "collective self-mastery"; social autonomy, on the other hand, can be regarded as the microcosm or the miniature version of political sovereignty at micro social spheres.[70] Both the individuated political sovereignty and social autonomy were conceptualized in terms of the empowerment of the individual citizen to "fully express" his political wills and to have these wills "fully integrated" into political decision-making at various macro and micro levels in society. By endowing the individual with these powers, the "conception of democracy as the political empowerment of the citizen" takes liberal democracy to task through giving, not just formal expressions, but also real substance to the liberal-democratic notions of citizenship and individual rights. Moreover, given that these powers would be expressed directly by individual citizens, the "conception of democracy as the political empowerment of the citizen" collides head-on with the "liberal-democratic conception of democracy" that is premised on the idea of representative government.

Though the "conception of democracy as the political empowerment of the citizen" resurrects the Rousseauean idea of popular sovereignty as the criterion of democratic legitimacy, it discards the idea of the "general will" altogether. Instead, it works with the notion of the "collective will." As is argued in Chapter 3 of *The Betrayal of an Ideal*, the notion of the "general will" is a problematic construct. Not only is the notion ill-defined and obscure—and can lend itself to extreme interpretations such as the claim of being "always right"—but also the very process of its formation or discovery is a mystical and undemocratic one.[71] The idea of the "collective will," on the other hand, is clearly defined. A collective will is a collectively decided will that emerges from the process of conjoining the wills the citizens enter directly into the collective decision-making process, with the intention of collectively forming a public will on a given issue. A collective will that emerges in this process would have the status of a democratically made decision, for it presents the embodiment of the sovereignty of all (the people)—in that it has resulted from the amalgamation of the individuated sovereignties of each and every citizen (all citizens being free and equal). The claim to being democratic here is contingent upon the general agreement among the public that the amalgamation-composition scheme used in the process was a "reasonably fair and accurate" one. (The task of elaborating on this characterization of the process is left to Chapter 2.) Lastly, another consideration that enhances the legitimacy of the collective wills emerged in this process is that the collective wills in question are "compositions" that result from amalgamating morally and rationally reflected-upon wills of a civic-minded and politically knowledgeable citizen body.[72]

Moreover, the idea of the "collective will" here differs from the Rousseauean "general will" in one more respect. Being merely an expression of the *common interests*, the general will ignores the particular interests of the citizens. In other words, it suppresses their private wills. And in doing so, it dissolves the individual in "the people" without acknowledging that this suppression disfigures or distorts the individual's conception of common interests. The collective will, on the other hand, though it has common or generalizable interests as its main object, allows some considerations of the private wills. (The discussion of the extent to which the collective will could incorporate the private wills of individuals is left to Chapter 2.) In conjunction with the micro principle of social autonomy, this feature of the individuated political sovereignty confronts the Rousseauean abstraction of the common interests, and at the same time, restores to the individual (a social being with both private and public wills) his fullness. It addresses the individual's need to express and impose his private wills side by side with his public wills.

In this respect, the "conception of democracy as the political empowerment of the citizen" also supersedes theories of deliberative democracy, all of which are premised on persuading the individual to suppress his private wills for the greater good of the community.[73] Despite this major difference with theories of deliberative democracy, the "conception of democracy as the political empowerment of the citizen" does require a comprehensive set of institutions that would be created solely for the purpose of public deliberations and civic education. However, one should be quick to point out that the "conception of democracy as the political empowerment of the citizen" assigns different roles to these institutions than the ones assigned by the theories of deliberative democracy. Rather than engaging citizens in deliberation on issues in order to develop consensi that would be reported to the public opinion poll-takers, or to the professional legislators and policy-makers, the deliberative institutions here are intended primarily as public institutions for political education and as forums for nurturing public-spiritedness and communal bonding. In their capacity as educational entities, these institutions would serve as public forums in which citizens would come to express their public wills, exchange ideas, and learn from each other, all of which would be geared and directed toward the goal of formulating their public wills. As was argued, the political decision-making power in the "conception of democracy as the political empowerment of the citizen" rests not in the deliberative institutions, nor in the hands of the professional legislators and policy makers who would take the will of the people as their guidelines in legislating the laws, but rather *directly in the hands of the individual citizens themselves*. In this conception, the people (i.e., the sum total of the individual citizens) are the sovereign power in the state, not just because the force of their collective wills guides the state, but mainly because *they themselves, and in a direct manner, create the collective wills that constitute the state.*

Finally, in returning to the micro principle of social autonomy, it should be added that without a doubt, the main domain of applicability for this principle is the workplace. At this level, the principle of social autonomy has much affinity with some familiar ideas of workplace democracy such as "industrial democracy," "economic democracy," "workers' democracy," and "workers' self-management." These ideas have been around for at least a century or so, and have received considerable attention in recent years.[74] However, given that this work is primarily concerned with the question of democracy at the macro levels of the state and national governments, a full discussion of the question of democracy at the workplace will not be provided here. Instead, the question will be taken up only when it would be relevant to the question of democracy at the macro levels. (The discussion of the question of property relations in Chapter 4 will be one such occasion.)

Notes

1. Theories of deliberative democracy are discussed in Chapter 13 of *The Betrayal of an Ideal*. See note 8 in the Introduction to this volume for a brief summary of the aims of these theories.

2. See note 1 in the Introduction to this volume for a concise statement on the ideals of democracy and the values they entail.

3. C. B. Macpherson's *Real World of Democracy*, *The Life and Times of Liberal Democracy*, and *Democratic Theory: Essays in Retrieval* have served as main sources of inspiration for these commitments.

4. Theories of participatory democracy are discussed in Chapter 12 of *The Betrayal of an Ideal.* See note 9 in the Introduction to this volume for a brief statement on these theories.

5. The "liberal-democratic conception of democracy" was introduced in Part II of *The Betrayal of an Ideal.* See note 2 in the Introduction to this volume for a brief statement on the conception.

6. See Chapter 11 of *The Betrayal of an Ideal* for a general discussion of the state of contemporary political philosophy, and its reluctance to take up the traditional question of sovereignty.

7. Two notes are in order here. First, in many respects, these two principles capture Dewey's contention that "[t]he idea of democracy is a wider and fuller idea than can be exemplified in the state even at its best. To be realized it must affect all modes of human association, the family, the school, the industry, religion" (Dewey 1954[1927], p.143). Dewey differentiates between "political democracy as a system of government" and "democracy as a social idea" (ibid.). While the general idea of the macro principle of political sovereignty corresponds to Dewey's notion of "political democracy," the micro principle of "social autonomy" directly ties with his notion of "democracy as a social idea." For Dewey, in an ideal political democracy, "the community itself shares in selecting its governors and *determining* their policies" (ibid., p.146, italics added). On the other hand, "democracy as a social idea," for Dewey, "consists in having a responsible share according to capacity in forming and directing the activities of the groups to which one belongs and in practicing according to need in the values which the groups sustain" (ibid., p.147). Now the second note. Here, the term "economic society," defined broadly, denotes the domain of economic activities, institutions, and entities that include, among others, the relations governing production and distribution, "means of production," markets, financial institutions, economic firms and corporations, and the activity of work itself. The "economic society" is distinguished from the "civil society" and the "political society." The term "civil society" denotes the sum total of all societal entities and spheres of activities that deal with, or relate to, the question of life in the society. Simply put, civil society is the sphere of life outside of the workplace, as the "economic society" is the sphere of life in the workplace, and in this sense, it is separate and distinct from the latter society. The "civil society" is also apart and distinct from the state (the "political society"). These domains will be defined and discussed in detail in Chapter 4.

8. Cunningham (1987), p.25.

9. See Chapter 3 of *The Betrayal of an Ideal* for a discussion of a positive interpretation of Rousseau's notion of sovereignty. According to this interpretation, sovereignty is the idea that the people ought to be the authors of the laws they abide by. Moreover, sovereignty is the collective exercise of the power of decision-making by the people on matters of generalizable interests, and in accordance with their political-moral ideals. Finally, sovereignty cannot be represented. See also note 71 below.

10. The term "*methodological* atomic individualism" is borrowed from Levine (1993), pp.26-27 and p.29. Levine distinguishes this type of individualism in method from "*substantive* atomic individualism" or "methodological individualism," "the view that individuals are radically independent centers of volition," or "according to which social facts are explained by facts about individuals, not social groups" (ibid.).

11. Rousseau's claim of the indivisibility of sovereignty is discussed in Chapter 3 of *The Betrayal of an Ideal.* (Also see Rousseau (1987[1762]), pp.30-31.) Here, Montesquieu and Locke come to mind even though Rousseau does not mention them by name.

12. " Man . . . is subject only to his own, yet universal, legislation, and that he is only bound to act in accordance with his own will which is, however, designed by nature to be a will giving universal laws . . . This principle, I will call the principle of *autonomy* of the will" (Kant 1959[1785], p.51).

13. See Chapter 2 of the *Betrayal of an Ideal* for further details.

14. This is in a sense a restatement of Macpherson's claim that the goal of democracy is "to provide the conditions for the full and free development of the essential human capacities of all the members of the society" (Macpherson 1965, p.37). Those readers who are inclined to take issue with this justificatory argument need to read on for, the "conception of democracy as the political empowerment of the citizen" can be justified in terms of other principles.

15. As is discussed in Part III of *The Betrayal of an Ideal,* Macpherson characterizes this conception of human essence as "developmental" (Macpherson 1973, pp.32-37, p.40, and p.52). He also characterizes it as an "ethical" and "democratic" conception of human nature. As Macpherson himself acknowledges, this conception of human essence is "fundamentally" a Marxian conception (ibid., p.36). Marx's conception, in turn, as Allen Wood puts it, is "profoundly Aristotelian" (Wood 1984, p.23). "For both philosophers [Marx and Aristotle] a fulfilling human life consists in the development and exercise of our essentially human capacities in a life of activity suited to our nature" (ibid.). While Aristotle's conception is tied to his notion of the good, Marx's is internally connected with a conception of communist society, as is clearly present in *The German Ideology* and other works.

16. This conceptualization of democracy as political institutions and social policies for maximizing positive human development stems from an "idealist" conception of politics that can be stated as follows. Politics, despite what it has been in the past or what it is now, ought to be the societal sphere and practices that have the social-legal realization of morally valuable ends as their ultimate goal. These ends are historically developed. That is to say, they are the ends that have withstood the test of history and have come to be recognized by general consensus of the past and present generations as ends desirable and valuable in themselves. In this sense, not only are these ends "universal," but also "intuitively" and "objectively" valid. Reverence and compassion for life, belief in and respect for the dignity and worth of each individual, individual autonomy, commitment to justice, amelioration of life's suffering, pacification and improvement of the conditions of human existence, and the positive development of the human individual (artistic, intellectual, moral, as well as physical) are among these ends. (In addition to being historically developed, one can further argue that these ends are also derivable in the Rawlsian original position.)

17. Pateman's conception is a case in point here. Elster also includes J. S. Mill in this category. See Elster's "The Market and the Forum: Three Varieties of Political Theory" in Elster and Hylland (1986). According to Elster, in this justification, "the goal of politics is the transformation and education of the participants. Politics, in this view, is an end in itself—indeed many have argued that it represents the good life for man" (ibid., p.103).

18. "Community," as Lesley Jacobs explains, is the contemporary gender-neutral substitute for the traditional ideal of "fraternity" (Jacobs 1997, p. 103). The term "community" is also intended here to stand for the "common good" or the "good of the community."

19. As is the case with *The Betrayal of an Ideal,* by liberalism, here it is meant a broad doctrine or a social-political world outlook that, taken from a narrow "historical" perspective, revolves around the advocacy of individual liberties as the primary social

value—and not as a spectrum of social-political views and positions that are often contrasted with "conservatism" and "libertarianism" in the contemporary politics of American liberal-democracy. When Ronald Dworkin takes "a certain conception of equality" as being the "nerve of liberalism," he takes liberalism in its contemporary sense in the United States (Dworkin 1985, p.183). Moreover, as is argued in note 7 of Chapter 6 in *The Betrayal of an Ideal*, no matter how much a liberal emphasizes the importance of equality, her emphasis should be understood within the context of, to use Waldron's phraseology, a "reference to a conviction about the importance of liberty (for everyone)" (Waldron 1993, pp.38-39). Rawls is a case in point here. Though he contends that justice (social and economic equality) is the "first virtue of social institutions," he goes on to posit liberty "lexically" prior to equality.

20. Individuals of this society are endowed with a package of liberties and with the equal right of access to them. Here this package is taken as a set of pre-given political and historical facts, as representing one of the highest political achievements of human civilization, enjoying the approval of the overwhelming majority (if not all) of the individuals in present-day American society, thus reflecting its tradition (or its "facticity" in Habermasian terms), as well as being morally justifiable in the Kantian sense of respect for persons. Finally, this package of liberties and the equal right of access to them is derived in the Rawlsian original position.

21. It is taken for granted here that the values in the trio (liberty, equality, and community), "under a reasonable set of definitions," as some have argued, are "mutually supporting" (DeMarco and Richmond 1986, p.12) and "harmonious and reconcilable" with one another (Arthur 1986, p.13).

22. Mill (1958 [1861]), p.130. This is how J. S. Mill explains a main benefit of "participation in political discussion and collective political action."

23. Ibid., p.54. J. S. Mill makes this comment in his exaltation of the Athenian democracy.

24. The important role played by Rousseau in some of the theories of participatory democracy is discussed in Chapters 3, 12, and 13 of *The Betrayal of an Ideal*. For instance, Pateman's theory was in part focused on the importance Rousseau placed on the educational value of participation.

25. Mill (1958 [1861]), p.25.

26. One should not lose sight of the problematic aspects of Mill's understanding of the question of democracy, which is discussed in some detail in Part II of *The Betrayal of an Ideal*.

27. Dewey also regarded democracy as "a way of life" and emphasized that it "cannot . . . depend upon or be expressed in political institutions alone" (quoted by Macpherson 1977, p.75). In Macpherson's interpretation, Dewey believed that this entailed the socialization of the forces of production and "a socialized economy" (ibid., pp.73-75).

28. Putnam (1992), p.180.

29. As is mentioned in Part II of *The Betrayal of an Ideal,* Tocqueville admired and idealized the early to mid-nineteenth-century practice of participatory democracy in the American New England states, especially in Connecticut and Massachusetts.

30. The other main justification for liberal-democracy is the consent justification—(that the people are to be ruled by their consent). Moreover, as can be recalled from the earlier part of this chapter, the "conception of democracy as the political empowerment of the citizen" takes the availability and protection of civil liberties and political rights for granted.

31. The community is the source of laws in the "conception of democracy as the political empowerment of the citizen," in that the members comprising the community jointly make its laws. It is through the direct participation of the members that the general interests of the community are best promoted. In his exaltation of the Athenian experience, J. S. Mill makes a strong argument for this claim. When the "public spirit" of Athens is absent, according to Mill, "[t]here is no unselfish sentiment of identification with the public. Every thought or feeling, either of interest or of duty, is absorbed in the individual and in the family. The man never thinks of any collective interest, of any objects to be pursued jointly with others, but only in competition with them, and in some measure at their expense . . . Thus even private morality suffers, while public is actually extinct" (Mill 1958 [1861], p.55).

32. In fact, Rodney Peffer has already made this argument for a participatory conception of democracy. As a way of improving Rawls' two principles of justice, "Peffer," according to Howard, "recommends adding to the equal opportunity principle 'an equal right to participate in all social decision-making processes within institutions of which one is a part'" (Howard 2000, p.8).

33. Marxist and Leninist conceptions of democracy are discussed in detail in Chapters 4 and 5 of *The Betrayal of an Ideal.*

34. Levine (1993), p.29. See note 10 above.

35. Macpherson has added the characterizations of "utility maximizer" and "possessive" to the individual in the liberal conception.

36. As will be suggested in the following section, the "common" interests or the "common good" of the community can also be regarded as the interests of the individual herself, the difference being that they represent her long-term and broad interests vis-à-vis her more immediate, direct, and short-term interests, often characterized as her "private interests." Moreover, in some respects, this notion is similar to Rawls' idealization of the citizens as reasonable and rational persons in *Political Liberalism.*

37. The Classical Marxism of Kautsky is a good example of the theoretical subjugation of the individual to community, whereas the Soviet Marxism is a good example for her theoretical-political subjugation. All of these conceptions take Marx's Thesis VI on Feuerbach—that human essence is "the ensemble of the social relations"—as their point of departure. It is relevant to note at this point that some critics of Marxism argue that Thesis VI is at the root of the "problems" of Marxism with democracy. For instance, according to Femia, "Marx's belief that human beings are entirely constituted by their social relations" forms the "genetic link" between Marxism and the despotic regimes that claimed to embody its ideals (Femia 1993, p.176).

38. The compromise conception being developed here owes much to Carol Gould's conception of the individual. It takes as its starting point Gould's postulation of individuals as "ontologically primary," as well as taking her notion of "individuals-in-relations" (Gould 1988, p.105). But, it soon supersedes Gould's by integrating the element of social-consciousness into the conception. Gould introduces the notion of the ontological primacy of the individual as a weapon to guard the individual against the theoretical onslaught of what she terms "holistic socialism" (ibid., p.100) that "disregards the ontological status of individuality" altogether (ibid., p.102). For her, the ontological primacy of the individual implies that "groups, social institutions, social structures, processes, and practices, as well as society as a totality, are derived entities that are constituted by the activities of, and relations among such individuals" (ibid., p.105). Postulating individuals as "individuals-in-relation" is intended to remedy the shortcomings of the conception of the individual in "liberal individualism" and in "pluralist political democracy." Gould

emphasizes that "relations among individuals are also essential aspects of their being" (ibid., p.105). She rightly stresses the importance of the relations among individuals, but falls short in emphasizing the role of consciousness in her conception of the individual. The compromise conception offered here attempts to overcome this shortcoming of Gould's conception by emphasizing the role of social consciousness.

39. In a sense, this interrelation between these two aspects of what constitutes the individual is analogous to the Freudian notion of ego—a compromise/negotiation between the id and the superego.

40. The Rawlsian "reasonable" person "is not the altruistic . . . nor is it [motivated by] the concern for self" (Rawls 1993, p.54). In the Rawlsian "reasonable society," "all have their own rational ends they hope to advance, and all stand ready to propose fair terms that others may reasonably be expected to accept, so that all may benefit and improve on what every one can do on their own. This reasonable society is neither a society of saints nor a society of the self-centered" (ibid.). Rawls' conception of the "reasonable persons" is directly tied to his fundamental idea of society as a fair system of cooperation among free and equal individuals. Rawlsian reasonable persons "are not moved by the general good as such" (ibid., p.50). What make Rawlsian individuals reasonable is their desire for a fair system of social cooperation: they "desire for its own sake a social world in which they, as free and equal, can cooperate with others on terms all can accept" (ibid.). They are capable of being persuaded to "discuss the fair terms that others propose" because they, as knowledgeable individuals, understand the importance of their sociality, as indicated by their desire for social cooperation (ibid., p.49).

41. Tocqueville attributed this doctrine to his idealized Americans of the early nineteenth century. "A man understands the influence that the well-being of the country has on his own; he knows that the law permits him to contribute to producing this well-being, and he interests himself in the prosperity of his country at first as a thing that is useful to him, and afterwards as his own work" (Tocqueville 2000, Vol.1, p.225). Tocqueville thought that the American "moralists" had already taught this principle to the American public and that it had been "universally accepted" by them: "[not] that one must sacrifice oneself to those like oneself because it is great to do it; but . . . [because] . . . such sacrifices are as necessary to the one who imposes them on himself as to the one who profits from them" (ibid., Vol.2, p.501). Although it seems that Tocqueville believed Americans had accepted this doctrine mainly on *pragmatic* grounds, at times one is led to believe that he thought Americans had also adopted it as a *moral commitment* as well as an *enlightened belief* (ibid., pp.500-06).

42. According to Macpherson, the descriptive model is relatively new to Western civilization and has increasingly gained prevalence only since the seventeenth century and fits perfectly well the worldview of the capitalist market society (Macpherson 1973, pp.9-10, pp.320-40). The attributes listed for ethical conception are to be found in Macpherson (1973), pp.32-37.

43. Macpherson (1973), pp.32-37. Also see Lindsay (1996), p.22.

44. The argument that espousing an essentialist conception of human nature does not obligate one to reject existentialist conceptions has been made persuasively by Peter Lindsay in Chapter 2 of his *Creative Individualism* (1996). The chapter in question is essentially a defense of Macpherson's "vision" of human nature. According to Lindsay, Macpherson's conception is "both ontological and historical," notwithstanding the fact that Macpherson does not make an explicit argument for it (Lindsay 1996, p.26). Lindsay defends Macpherson's vision against Nietzsche's "anti-ontology" and Sartre's claim that "there is no human nature" (ibid., p.16). In Lindsay's view, Macpherson's "vision" of

human nature takes Man as a "socially constituted, ontological and historical being," a view that he finds tenable (ibid., p.34).

45. Macpherson (1973), p.34. According to Macpherson, liberal democracy has already achieved such a synthesis. But, in his view, the problem with the way it has been done is that the notion of "utility maximizing" is predicated on a "system of market incentives and market morality" (ibid.). The synthesis preferred by Macpherson himself regards the utility maximizing as a "means" in the service of maximizing human essential powers (ibid.).

46. Gould (1988), pp.290-93.

47. One should add, the claim that public interests or the "common good" can be seen as the broad and long-term private interests of the citizens does not necessarily commit one to the proposition that every element of, and every aspect in, what is considered as the public good can be regarded as an aggregate of the private interests of the individuals forming the public.

48. The absence of "physical coercion" constitutes the main component of Isaiah Berlin's notion of "negative freedom." Moreover, the phrase quoted above is part of Macpherson's argument against the deficiencies of Berlin's division of liberties into "positive" and "negative" (Macpherson 1973, p. 117).

49. Rawls (1971), p. 303.

50. Gould (1988), p.257.

51. The modified version of Rawls' principles proposed by Rodney Peffer would be a more suitable set of principles of justice for the "conception of democracy as the political empowerment of the citizen." This is in part due to the fact that Peffer's principle 3(b) is similar to the "micro principle of social autonomy of the individual" presented earlier in this chapter. (See Peffer 1990, p.418 or Howard 2000, pp.32-33 for a statement of Peffer's modified version of Rawls' principles.) Concisely stated, Peffer modifies Rawls' principles by adding two new elements. First, Peffer introduces a new first principle (viz., "Everyone's security and subsistence rights shall be respected") and puts it lexically prior to Rawls' first principle. Second, Peffer adds his principle "3(b)" which states "There is to be . . . an equal right to participate in all social decision-making processes within institutions of which one is a part." One should add that Rawls has expressed his approval of Peffer's principle on meeting subsistence needs of the individuals, stating that it is implicit in his own first principle; however he has disagreed with principle 3(b) on the grounds that he "should not include its being required in the first principles of political justice" (Rawls 1993, p.7n7).

52. Habermas (1996), p.59.

53. As was mentioned in the preceding section, the notion of "democratic personality" is borrowed from Carol Gould and it includes traits such as tolerance, "flexibility and open-mindedness," "commitment and responsibility" (Gould 1988, pp.290-93).

54. This question will be revisited and discussed in some detail in Chapter 4.

55. Two notes are in order here. First, this approach to the question of rights affirms the characterization of the "conception of democracy as a political empowerment of the citizen" as an *historical* and concrete formulation of the question of democracy, as was done in the opening pages of this chapter. Second, the advances in the technical-material state of society increases our sense of what we are entitled to (and how much). For example, advances in the socio-economic levels of society give rise to the discourse of entitlement or rights to unemployment benefits or healthcare.

56. Rawls (1985), p. 246.

57. As was stated in the Introduction to this volume, throughout this book the term "will" is used in a technical sense. A "will" denotes a "public judgement." That is to say, a particular sort of opinion on a given issue that; a) is well-informed and educated on the issue and has the knowledge of alternatives, b) is subjected to reflection on the basis of moral considerations and reasonableness, c) is preferably scrutinized in public deliberations, and d) has the "force of the commitment" of the individual citizen who holds it. As will become apparent in the rest of this chapter and Chapter 2, an individual citizen may have more than one "will" on any given public issue.

58. One should recall that sovereignty was defined as the power of the people to make political decisions or legislate laws.

59. One can further argue that a "collective will" formed in this manner should also be regarded as representing the will of each and every participant, regardless of whether or not the formed collective will corresponds to what a given participant contributed to the process of its formation (even if the resulting "collective will" turns out to be the opposite of what the individual brought into the process as her own expression of the public will on the issue in question). This latter argument is mainly based on the contention that the individual does participate in the formation of the collective will in question, and she does so as a free and equal member of the society. This argument also follows from the contention that the scheme does allow her to fully express her wills and does fully incorporate these wills of hers into the decision-making process, all of which point to regarding her as a co-author of the formed collective will. Admittedly, this argument is weak, and thus needs to be bolstered. This will be done in some detail in Chapter 2. For the time being, it suffices to say that this argument is mainly intended to persuade the individual to perceive the collective will or decision produced by the "idealized" scheme less as an obligation to abide by, and more as a decision or law co-created by her. (This applies to cases in which the produced will or decision turns out different than, or the opposite of, the wills the individual entered into the scheme.) Regarded as such, this argument is related to the second argument above which was mainly intended to address the question of the democratic legitimacy of the collective wills or decisions produced by the scheme. Moreover, as will be argued in Chapter 2, this argument does not amount to contending that the individual is morally obligated to abide by the decisions produced by the "idealized" scheme; nor can it be construed as claiming that it resolves the problem of the state authority vs. individual autonomy.

60. No doubt, being morally persuaded is *not* tantamount to being morally obligated. (This question will be discussed in Chapter 2.) Moreover, in a similar fashion, Kymlicka argues that "[m]any studies have shown that citizens will accept the legitimacy of collective decisions that go against them, but only if they think their arguments and reasons have been given a fair hearing, and that others have taken seriously what they have to say" (Kymlicka 2002, p.291).

61. Another way to underscore the differences between the individual's "public wills," on the one hand, and her mere preferences or opinions, on the other hand, is to use Daniel Yankelovich's distinction between the individual's "public judgements" and her "top-of-the-mind, offhand views," which are expressed in "public opinions." The following description of the "public judgement" adequately defines what is meant above by the "public will." According to Yankelovich, "public judgement" means "a particular form of public opinion that exhibits 1) more thoughtfulness, more weighing of alternatives, more genuine engagement with the issue, more taking into account a wide variety of factors than ordinary public opinion as measured in opinion polls, and 2) more emphasis on the

normative, valuing, ethical side of questions than on the factual, informational side" (Yankelovich 1991, p.5).

62. The concept of "general will" is discussed in Chapter 3 of *The Betrayal of an Ideal*. It is argued there that the "general will" is an ill-constructed and problematic notion and lends itself to extreme interpretations such as the claim that it is "always right." See also note 71 below.

63. Tocqueville's "doctrine of self-interest well understood" was used earlier mainly in support of the argument that pragmatic considerations could motivate the individual to curb her private interests in order to serve the public interests. (See note 41 above.) This is how Tocqueville described his idealized American characters: "the inhabitants of the United States almost always know how to combine their own well-being with that of their fellow citizens" (Tocqueville 2000, Vol.2, p.501). And that "they complacently show how the enlightened love of themselves constantly brings them to aid each other and disposes them willingly to sacrifice a part of their time and their wealth to the good of the state" (ibid., p.502).

64. Sunstein (2000a), pp.114-15, italics added.

65. In somewhat analogous way to Sunstein's argument—to the effect that empirical evidence suggest that individuals qua citizens behave differently than individuals qua economic consumers—Philip Pettit argues that individuals as social agents do not behave as self-regarding economic rational choice-makers in their interactions with each other, and that the economic model of the individual as "self-centered" or "self-regarding" is in conflict with empirical evidence, or as he puts it, with "common sense" (Pettit 2001, pp.81-82).

66. Even in those rare cases when the individual gets an opportunity to express his wills, he does not get to do it "fully." He has to express his wills monosyllabically (by saying a definite yes or a definite no).

67. This characterization of social autonomy distances the idea from what Weale calls the "strong" sense of autonomy found in the Kantian tradition: "the idea that persons are by their nature self-governing creatures, that is beings whose moral personality finds its fulfillment in their prescribing principles of action to themselves" (Weale 1999, p.64). This characterization also distances social autonomy from the "weaker" sense of autonomy which takes it as "the non-paternalist proposition that people are the best judges of their own interests" (ibid., p.65). Although social autonomy is concerned primarily with the private interests of the individual and the small social domains, the interests in question are, to use Joseph Raz's terminology, "inherently public goods" or "collective goods" (Raz 1986, p.199). That is to say, these interests in essence boil down to the question of the empowerment of the citizen or the social units that are compatible with the same degree and scope of empowerments for other individuals or social units.

68. In many respects, the notion of social autonomy is a miniaturized version of David Held's notion of the "democratic autonomy" of the individual in a democratic community. According to Held,

> [t]he principle of autonomy, entrenched in democratic public law, ought to be regarded . . . not as an individualistic principle of self-determination, where "the self" is the isolated individual acting alone in his or her interests, but, rather, as a structural principle of self-determination where "the self" is part of the collectivity . . . enabled and constrained by the rules and procedures of democratic life. Autonomy, in this account, has to be understood in relation to a com-

> plex base of rules and resources—individuals are equally free when they can enjoy a common structure of political action. Hence, this form of autonomy can be referred to as "*democratic autonomy*"—and entitlement to autonomy within constraints of community. It can be clearly distinguished from unbridled license for the pursuit of individual interests in public affairs. (Held 1995, p.156, italics added).

69. Some of the problems that can haunt deliberation sessions, including Rousseau's reasons for banning them, are discussed in Chapter 13 of *The Betrayal of an Ideal*. The phrase inside the quotation marks belong to Rousseau.

70. Phrases inside the quotation marks are borrowed from Levine (1993), pp.78-79. Levine's definition of democracy is similar to the "conception of democracy as the political empowerment of the citizen" in some respects.

71. See Chapter 3 of *The Betrayal of an Ideal* for a detailed discussion of the "general will." Briefly stated, the discussion rejects Rousseau's extreme interpretation of what the concept represents and, at the same time, rejects Rousseau's understanding of how the will is assumed to be revealed. This in turn requires abandoning altogether the claim that the general will exists as an objective "matter of fact" waiting to be revealed and, at the same time, it necessitates the qualification of the claim that "the general will is always right." This gives way to a thinner notion of the "general will" that treats it primarily as a will that stands for the common or generalizable interests of the community and its ethical ideals. Moreover, this thin notion takes the general will primarily as a "popular will" that is generally willed by a competent citizen body. Viewed as such, the "general will" would draw its political legitimacy or its claim to "righteousness," not from the claim that it is objectively valid or rational, but from the empirically verifiable claim that it is generally (i.e., popularly) willed by a citizen body composed of civic-minded members. This interpretation, opens the door to the possibility that the general will could also lay claim to being rational, or at least, to being "reasonable." See also note 9 above.

72. One should recall that the wills entered by citizens into the "idealized" decision-making process are influenced and informed by free and reasoned public deliberations that shed light on the issues in question and educate the citizens on the differing interpretations of the common good related to those issues.

73. See Chapter 13 of *The Betrayal of an Ideal* for a discussion of this point.

74. G. D. H. Cole's works during the years 1915-1920 are perhaps the most significant of the original contributions to the idea of industrial democracy in the early part of the twentieth century. Pateman's *Participation and Democratic Theory* (1970) relied heavily on Cole's works. Among the most recent works, one should mention *After Capitalism: From Managerialism to Workplace Democracy* (Melman 2001), *Self-Management and the Crisis of Socialism* (Howard 2000), *The Practice of Workplace Participation* (Denning 1998), *Empowerment and Democracy in the Workplace* (Dew 1997), *Economic Democracy: The Politics of Feasible Socialism* (Archer 1996), *Democracy and Efficiency in the Economic Enterprise* (Pagano and Rowthorn 1996), *Restoring Real Representation* (Grady 1993, chapters seven and eight), *Power and Empowerment: A Radical Theory of Participatory Democracy* (Bachrach and Botwinick 1992), *When Workers Decide: Workplace Democracy Takes Root in North America* (Krimerman and Lindenfeld (editors) 1992), *From the Ground Up: Essays on Grassroots and Workplace Democracy* (Benello 1992), and *Workplace Democracy: The Political Effects of Participation* (Greenberg 1986). The reader should consult note 61 in Chapter 3 for a list of titles on democracy at the local and community levels.

Chapter 2

A Different Kind of Collective Decision-Making Scheme

Chapter 1 conceptualized democracy as the idea of the political empowerment of the citizen. For macro levels, this conceptualization developed the notion of the "individuated sovereignty" and envisioned an "idealized" scheme of collective decision-making that was intended to give a conceptual expression to exerting this sovereignty *directly*. For micro levels, on the other hand, the "conception of democracy as the political empowerment of the citizen" put forth the notion of "social autonomy" that was intended to empower the individual to exercise some degree of control over her circumstances in the social milieu of her everyday life, e.g., the workplace. Although the "idealized" scheme of collective decision-making was devised primarily for the macro levels, as was suggested, it could also be modified so that it would serve a similar purpose at micro levels.

As a way of giving a practical expression to the "conception of democracy as the political empowerment of the citizen," in general, and to the ideal of the direct exercise of the "individuated sovereignty" in particular, this chapter will attempt to develop a scheme of voting that could be regarded as the actualization of the "idealized" scheme of collective decision-making introduced in Chapter 1. There, the idealized scheme was premised on the assumption that the individual citizen would possess two types of powers: the power to "Fully Express" her public and private wills, and the power to "Fully Integrate" these wills into the process of producing collective decisions, which were referred to as "collective wills."[1] The voting scheme to be developed in this chapter (which will be referred to as FEFI), will work with these powers.[2] The process of developing the scheme in question will begin by presenting first a simple system of voting that gives citizens virtually infinitely many different ways of expressing their wills, and then incorporates these expressions as inputs into an amalgamation-composition process that produces collective wills. This presentation will then be followed by a discussion of a number of potential criticisms that could be directed against the scheme. Based on this discussion, the scheme will then be further developed into a more complex system that would be able to deflect most criticisms. What follows is the description of the simple scheme.

A Different Kind of Voting System

It is possible to devise a collective decision-making scheme that would offer to citizens multiple policy options and give them the opportunity to express their wills for one or more of these options in the fullest feasible way possible. As an example, suppose that citizens are given the opportunity to make the decision on the question of how the nation should plan its budget spending for the upcoming year. Further, suppose that they are presented with a list of six meaningful and varied policy options on the issue (options I-VI). Finally, suppose that they would be making their decisions by expressing their wills in three categories or layers of "private-interest," "particular-public-interest," and "general-public-interest." Each citizen would be given 100 points to vote with in each category (a total of 300 points for each voter). Nellie, a citizen, would have the choice of spreading her 100 points among the six budget policy options in any way she believes would express her wills accurately or fully on the issue. In spreading her 100 points among the budget policy options in the "private-interest" category, she would think in terms of which budget option(s) would serve her private-interests best. She might end up deciding that budget option III is the best of the six options, yet options II and V have some merits as well. As a result, she would divide up her 100 points into three parts. She would place, let us say, 64 points on budget option III, 24 points on option II, and the remaining 12 points on option V—(this would mean that no points would be placed on options I, IV, and VI).[3] In voting in the "particular-public-interest" category, she would think along the lines of deciding which budget option(s) would be better for the nation in the short run and in terms of its most immediate needs, and would spread her 100 points among those options which, in her judgement, would be most beneficial to the public. It might well turn out that in voting in this category, she would spread her points on the same option she favored in the first category, but in a different proportion; or that she would place some of her points on options that she deemed unacceptable in voting in the private-interest category.[4] Finally, in spreading her points in the "general-public-interest" category, she would judge which budget option(s) would benefit the public's general interests in the larger scheme of things, and in the long run, and place her points on those options in accordance with her judgement.[5] Once done with spreading her voting points, Nellie would walk away from the voting booth thinking that *she has fully expressed her wills*. (One should note that in this scheme, the voter has virtually infinitely many different ways of expressing her wills on the issue.[6])

The next step in this scheme (FEFI) is the process of the full integration of the wills expressed by all citizens who took part in the decision-making. This process is simply a matter of tallying the points each policy option has accumulated. This can be performed by a computer that receives these points as data, and finds the arithmetic sum of the points each policy option has received. If a policy option receives a certain prescribed qualified majority (e.g., the absolute majority, 50%+1 point), that option will be regarded as the winning policy option, or putting it differently, as the collectively decided policy choice, or the formed "collective will." If no policy option emerges as the clear winner, then there would be a runoff vote between the two options that have received the most points.[7] (As one might suspect, this scheme of voting would require an extensive and fast network of powerful electronic media. A thorough discussion of this requirement will be taken up in Chapter 3.)

Moreover, FEFI would offer a "protest option" available to citizens as the seventh option to choose, in case they are dissatisfied with the range or meaningfulness of the six options presented to them on the ballot. If a citizen believes that the six policy options listed on the ballot fall short of offering her a meaningful and complete list options to choose from, she has the option of placing all of her 100 points in one, or two, or in all of the categories of private-interest, particular-public-interest, and general-public-interest on the seventh option in order to protest the ballot.[8] At the conclusion of voting, the combined protest points will be sent to the agenda setters as feedback to be studied for policy option development purposes in the future. The protest option can also function as a "veto vote," if it receives the largest share of points in the primary stage, or receives a certain specified percentage of points, e.g., 35 percent, or makes it to the runoff stage as the second highest point receiver and wins an absolute majority there. A "veto vote" is an indication that citizens did not approve of the policy options presented to them and the agenda setters must put together another set of policy options based on studying the results, and then schedule another public voting session.

In short, FEFI first quantifies the expressions of the wills of the voters and then amalgamates-composes a single will (a collective will) from these quantifications. It is of utmost importance to recall here that FEFI is intended to give an actual expression to the "idealized" scheme of collective decision-making developed in Chapter 1, and thus should be regarded as presupposing all of the theoretical assumptions of the conceptualization of democracy as the political empowerment of the citizen, including the democratic conceptions of the individual and society.

Having described the actual working of FEFI, it is now the time to consider how the collective wills produced by the scheme can be characterized as democratic. To begin with, one should note that a collective will formed by FEFI should be interpreted as being the strongest composition of wills among the wills expressed by voters, and thus, as the strongest expression of the voters' judgements on the issue put to vote. Alternately stated, the will produced by FEFI should be regarded as the strongest expression of the existing shared conceptions among voters on what ought to be done. In the Rousseauean lexicon, the collective will formed by FEFI could be regarded as a complex composition that results from the "general will" on the one hand, and the "will of all," on the other hand.[9] The democratic legitimacy of the collective will produced by FEFI is drawn not just from the fact that it has resulted from the direct participation of all on an equal basis, nor just from the fact that the participants have expressed their wills fully, and not just from the fact that their wills have been taken fully into account in the process of the will-formation; but, also from the fact that their wills are part and parcel of the collective will. That is, their wills have been interwoven into the fabric of the collective will. They themselves, and in a "hands-on" and direct manner, have created and authored the collective will. It is of their very own making. The collective will arrived at here is the will of the citizens; not just conceptually, virtually, figuratively, nor by means of some long or convoluted explanations and arguments; but rather *immediately, directly, literally, and actually.*

It should be added that these contentions offered on behalf of the democratic legitimacy of the collective wills formed by FEFI do not amount to arguing that the collective wills in question are necessarily the "right" collective wills (or decisions), or the "best" decisions, or the most "accurate" revealers of the popular or "dominant" wills of the voters—(these questions will be dealt with later). Rather, they make the case that the formed collective wills should be viewed as the ones that are obtained the "right" way and in the "best" possible way of obtaining them collectively and democratically, i.e., within a framework that gives all citizens *direct, full*, and *equal* power in the decision-making process, as well as giving them virtually an infinite number of choices in exercising this power. (These contentions will be elaborated further in this chapter. A fuller discussion of the question of democratic legitimacy will be taken up in Chapter 4).

The claim above, that citizens' wills are part and parcel of the "collective wills," rests on the contention that, since each citizen can express her wills for more than one option, and that she can do this in three categories, and in each category in varying ways and degrees; therefore, there exists a high probability that the formed collective will has received some points from each voter (unless the voter in question has placed all of her points in all of the categories on the policy issue(s) that lost out).[10] Now, in light of this probability, the age-old and paradoxical problem of compelling the (opinion) minority to obey the wills of the (opinion) majority loses its force and cogency.[11] Moreover, one should add that the notions of majority and minority, given that the decision was not arrived at by saying a definite yes or a definite no, lose their traditional meanings and become irrelevant to this process of decision-making. This, in part, results from the fact that FEFI blurs the line that divides minority and majority.

Furthermore, in light of the fact that all voters are given direct, equal, and full powers, as well as being given a virtually infinite number of choices in expressing *and* entering their wills into FEFI, and also given the other characteristics of FEFI and the collective wills produced by it—(as discussed just above, especially its democratic legitimacy)—one can further argue that voters would find themselves both morally and rationally persuaded to accept the authority of the collective wills produced by FEFI. The knowledge *and* the fact that the individual herself is an equal and direct co-author or co-creator of these wills, lighten the burden of complying with them. In the FEFI scheme, the individual does not see herself just as a *subject* of the decisions or wills produced, but also as part and parcel of the *authority* that produces them. Thus viewed, the dichotomous relation between the political authority, on the one hand, and the individual autonomy, on the other hand, appears less severe, and thus less problematic.[12] In the scheme of FEFI, by obeying the dictates of a collective will, the individual obeys herself.[13] However, this is not because she is the sole decision-maker in the process, or because the will in question has been decided unanimously, but because she has been a free, full, equal, and direct participant in the process of forming that will, which she regards as being "adequately democratic."[14]

To illustrate this latter point, one can liken FEFI to the process of deciding the champion in an ice hockey league, and selecting the members of the national ice hockey team from the players in the league at the conclusion of the season. The individual as a player on one of the teams in the league accepts the outcomes of the games he and his team have played, even the ones they have lost. He also acknowledges the championship status of the team that eventually wins the cup, and supports the national team, provided that there is a general consensus that the games were refereed fairly and the process was fair, and his team and he had pretty much the same opportunity (and resources) as others to perform to the best of their abilities. He accepts the final outcome of the season, notwithstanding the possibility that he might think he is a better player than some of the ones chosen for the national team or that his team is a better team (theoretically, i.e., mathematically, statistically, etc.) than the one that won the championship cup. He knows that he was a contributor to the process that led to the championship status of the team that won the cup, and the selection of the national team. He also knows that these outcomes are partially his own making, and he has directly contributed to the process, even though he himself is not part of either team. He supports the national team, even though he is not in it. In acknowledging the league's champion and the national team, he recognizes their importance and accepts the final outcome of the process. Given that the process is regarded as adequately fair and equal, he most likely will agree that the championship team was, at least, one of the better teams in the league, if not necessarily the best.

Now returning to FEFI, one can argue that, by the same analogy, the citizen whose prevailing wills (or most preferred policy option) did not prevail as the collective will, will concede that the prevailing collective will must be one of the better policy options. He might think that he is more knowledgeable and more (socially and politically) intelligent on the issue just decided than most of his fellow citizens; but, given that he is willing to grant that most of his fellow citizens have adequate levels of political intelligence and are knowledgeable about the issue, he will acknowledge that the formed collective will must be a good one (or at least not a bad one)—notwithstanding the possibility that he might still think that his prevailing wills (i.e., the wills he expressed by voting) were better than the one that ended up prevailing. In short, he will accept, submit to, and obey the authority of the formed collective will based on both moral and rational grounds.[15] It should be noted that this is in addition to the practical (instrumental) rationality for obeying the authority, according to which obeying laws is a prudent thing to do (for personal reasons), and also that doing so promotes social cooperation, and thus is indispensable to the sustenance of democracy and the social contract.

To complete the picture of FEFI, one also needs to include the following arguments. First, although FEFI is primarily intended to address the question of exercising the individuated sovereignty *directly* at macro levels, with some modifications, it can also be used as an instrument for the "full expression" and "full integration" of the individual's social autonomy at meso and micro levels. Second, in comparing FEFI to the existing voting practice of the American liberal democracy, what is markedly different about FEFI is that it makes it possible to exert the sovereignty of the people on an *ongoing* basis. Unlike what FEFI is intended to accomplish, citizens in liberal democracies surrender their sovereignty (their power to legislate laws) to their representatives via the mechanism of the periodic and intermittent elections, once every few years. As a result, the citizens are truly, actually, and directly sovereign only in those short-lived intervals or moments. The rest of the time, they are sovereign *only* in name; that is, indirectly and theoretically (i.e., in the final analysis and ultimately).[16] Third, FEFI should not be mistaken with what is commonly known as the "interest-aggregation" procedure; for what is being aggregated, or rather amalgamated-composed, here are *wills* and not (personal) interests or self-regarding preferences. True, personal interests do play a part in the formation of these wills, but only in a considerably lesser extent than the public interests do. However, despite this crucial difference, given that the scheme is, after all, a variant of aggregation schemes, one should expect that it would be subjected to the same sort of criticisms that are often leveled against the interest-aggregation procedures. Some of these relevant and potential criticisms will be considered later in this chapter.

Fourth, in line with the assumptions of Chapter 1, FEFI is premised on the existence of a politically knowledgeable citizenry and a set of public-deliberation and civic education institutions intended for the political education of the citizenry on the issues that are put to the public for vote. These institutions would provide ample opportunities and channels for citizens to learn about the issues in question and contribute to the learning of their fellow citizens, before public voting takes place. Moreover, these institutions are essential instruments in helping to transfer citizens' public or politically and morally based preferences to "public wills." Fifth, the distinction between the public or politically-morally based *preferences*, on the one hand, and public *wills*, on the other hand, is drawn on the contention that public wills are reflected-upon and considered judgements of the citizens and aim at the good of the community. Furthermore, public wills carry the force of the commitment of the citizens; that is to say, the willingness of the citizens to spend their time and energy to bring these wills to the attention of the public and present them as possible courses of action.[17] In this respect, public wills are *active* in nature and thus differ from mere opinions and desire-based or self-interested preferences that are essentially *passive*. Stated alternately, public wills are to be regarded as representing the expressions of the reasoned and politically-morally considered opinions of citizens, as well as being the expressions of their commitments to the well-being of their community and their judgements on what would be the best course of action for the community to follow on the issue put to vote. Finally, it is worth stressing at this point that FEFI is proposed against the backdrop of the democratic conceptions of society and the individual offered in Chapter 1.

* * * * *

Having explicated this collective decision-making scheme somewhat adequately, one should now anticipate and confront a number of potential criticisms that could be directed against it. First, it can be argued that FEFI is based on the naïve assumption that citizens would abide by the prescribed procedure for this voting scheme and make a sincere and honest attempt to distinguish between their private wills, on the one hand, and their public wills, on the other hand. Second, given that some citizens have strong feelings or religious or ideological attachments to some issues, they might disregard this distinction altogether and place all of their voting points on policy options that affirm their commitments to their views or beliefs—this will be referred to here as "ideological voting." Third, it can also be argued that given the fact that issues are often complex, citizens might not be able to make the suggested distinction between their private and public wills even if they sincerely attempt to do so. Fourth, the "truth" of the collective will arrived at can be questioned by arguing that outcomes can be manipulated by voters' "ideological voting" patterns or by their "strategic voting" schemes (i.e., "insincere" and "mischievous" voting), as well as by the strategic maneuvering and influence-peddling among agenda-setters and by those who would decide which agenda or issue to be put to the public vote.

Moreover, it can further be argued that the agenda setters can also manipulate the outcomes in a variety of other ways, such as by controlling the agenda through restricting it, or by the very ordering of the policy options to be presented to voters, or simply by the way they would phrase these options. Fifth and finally, it can also be argued that, even if these four criticisms could be deflected satisfactorily, the "correctness" or "truth," or "accuracy" of the collective wills reached via this scheme would still be open to question, for it is possible that for a given specific profile of wills, the outcome could be different if a different scheme of amalgamation-composition is used. One should note that the fourth and fifth contentions above are epistemological in nature, and skeptical for that matter, for they question whether a "popular will" or a "public will" could ever be truly revealed or known. Moreover, as will be seen later in this chapter, the fifth contention above hints at an ontological problem. That is to say, the skepticism about the knowablity of public wills can easily push one down a slippery slope leading to ontological skepticism, and thus to questioning the very existence of public wills and public interests as social objects.

The remainder of this chapter will be devoted to discussing the problems brought to light by these criticisms. As will be seen, FEFI will be further developed in order to deflect some of these criticisms. For the sake of organization, these problems will be divided into four categories as follows: (1) the problem of knowledgeable and competent voters, (2) the problem of "irresponsible," "ideological," and "strategic" voting, (3) the problem of the manipulation of the agenda, and (4) the problem of revealing the wills of the voters "truly."

The Problem of Knowledgeable and Competent Voters

One cannot stress enough that both the "conception of democracy as the political empowerment of the citizen" and FEFI—which is intended as an instrument of this empowerment—are premised on the existence of a politically knowledgeable, civic-minded, and intellectually competent citizen body. Democracy of the magnitude being envisaged here, no doubt, would not be possible or desirable without such a body politic. This requirement, as will be discussed in Chapter 3, could be met by instituting a system of political education that would give citizens ample opportunities to learn about the issues, and deliberate about them among themselves in various public deliberation forums. It should be emphasized that although the "conception of democracy as the political empowerment of the citizen" is essentially a will-composition conception, it also has a strong component of politics as dialogue interwoven into its fabric as a pre-requisite. This component will be discussed in further detail in Chapters 3 and 4. As will be seen, Chapter 3 will present a theoretical and institutional framework for considering a system of pre-voting deliberations that could be created exclusively for the purpose of the political education of the citizens. Chapter 4, on the other hand, will provide an overview of some of the prevalent skeptical views that hold the political knowledge and competence of ordinary citizens in a negative light, and will argue toward undermining their credibility.

The Problem of "Irresponsible," "Ideological," and "Strategic" Voting

In addressing the potential criticism that FEFI is based on a naïve assumption that citizens would make honest and sincere attempts to distinguish between their private and public wills, one should bear in mind that the present-day American liberal-democracy operates on a similar "public" assumption. That is to say, on the one hand, the democratic system trusts that the citizens, by and large, vote responsibly, and on the other hand, citizens, generally, "trust the system," so to speak. They trust that "the system works" and that their fellow citizens, by and large, vote "responsibly." That is to say, they vote more in accordance with their conceptions of the good of the community and less with their personal prejudices and narrow self-interests—except when they have high stakes and vital interests in the issues to be voted on, in which case they are justified in voting according to their self-interests. The voting scheme proposed here rests on this same assumption. It assumes that voters, by and large, live up to the expectations of citizenship conferred upon them by the democratic system, and by their fellow citizens; and that they trust that the system works. This assumption is a crucial one; for, democracy cannot function without a high quality of citizenship and mutual trust. It is in this light that FEFI assumes that citizens would make an honest effort to distinguish between their private and public wills, and assign higher weights to their public wills than to their private ones—the ratio of 200 points to 100 points.

It should be added here that one of the reasons for proposing a three-layered voting scheme is to "compel" voters to reflect upon what the issues entail and how each policy option, if passed, would affect their lives as well as the society as a whole. Moreover, one should also add, the assumption that citizens are capable of distinguishing between their private wills, on the one hand, and their public wills, on the other hand, is not an idealistic or naïve notion. The same can be said about the willingness of citizens to assign a greater weight to their public wills over their private ones. There exists ample evidence to suggest that the citizens of present-day society do make this distinction and do privilege their public wills over (and sometimes against) their private ones when they get the opportunity to speak their wills.[18]

The other side of the coin in voting "irresponsibly" is the case in which a citizen discards her citizenship responsibilities of thinking or behaving democratically and votes "ideologically." In voting in FEFI, the ideological voting pattern would be the case in which a voter places all of her points on a policy option that best affirms her religious or strongly held ideological beliefs. The problem of ideological voting is already present in existing voting schemes. The most well-known example here is the voting patterns of the Christian Coalition in the United States, whose voters vote along ideological lines in order to support religiously inclined candidates or policies. Without a doubt, there are considerable numbers of other individual voters who stand in direct opposition to the Christian Coalition and vote ideologically against it—often as a reaction. In this group, one can include the staunch feminists, gay-lesbian activists, environmentalists, and the leftists and left-leaning individuals and organizations. Although the Christian Coalition appears politically stronger than the rest of the ideological voters—mainly because of its financial and organizational strengths and because it votes as a bloc—there is no evidence to suggest that the number of votes the Christian Coalition musters exceed the number of votes cast by those individuals who vote with the intention of counteracting the effects of the votes of the Christian Coalition ideologically.[19]

As will be seen later in this section and in Chapter 3, the problem of ideological voting will not pose a serious difficulty to FEFI, for, in this scheme, the organizational and financial strengths of ideological groups would cease to be relevant. What would be relevant instead is the actual number of the ideological votes cast. It seems intuitively plausible to assume that in a sufficiently democratic culture—and in the absence of any organizationally or financially privileged group or class—substantial ideological strays from the democratic path would inspire ideological reactions with roughly equal intensity and magnitude. Thus, in FEFI, the influence of the opposing ideological votes would cancel each other out, in a manner of speaking, and the issues would end up being decided, by and large, by the rest of the citizens who would vote responsibly.

Closely related to the problem of irresponsible and ideological voting is the problem of voting "strategically." Granting assumptions that voters do, by and large, vote "responsibly" and that opposing ideological votes cancel each other out, does not make the problem of "strategic voting" go away. Strategic voting is an "ineradicable possibility" and one can only hope to minimize it by making it an unattractive option.[20] The questions of the extent to which citizens of present-day society vote "responsibly," and how widespread "strategic voting" is, and how accurately the actual votes cast represent the true wills of the citizens rather than their strategic wills, lie outside the scope of this discussion. Nonetheless, one can argue that there are cogent reasons to believe that engaging in strategic voting by citizens results mainly from the deficiencies of the existing voting schemes. Moreover, there are no compelling reasons to believe that strategic voting is a necessary evil, intrinsically associated with the enterprise of voting itself, let alone assume that it is widely practiced and that it distorts the voting outcomes significantly. Rather, what seems more compelling is to argue that strategic voting is a by-product of the deficiencies of the existing voting practices, which not only deny meaningful alternatives to voters, but also give them a deficient method of making choices or expressing their wills. Furthermore, one should also point out that the practice of strategic voting by voters, if ever it can be shown to be widespread, has been observed and studied only in relation to the elections of candidates running for office and not in referenda on issues. It could well be the case that the presence or prevalence of strategic voting is only associated with the election of persons. As usually is the case, these elections pit parties against one another, as well as pitting the factions inside the parties against one another. The phenomenon is closely related to factionalism, especially within parties and legislative bodies. The voting scheme proposed here, on the other hand, is intended for making public decisions on the issues only, and not for electing officials, nor for making decisions within the confines of legislatures.

Having mentioned these preliminary remarks about irresponsible, ideological, and strategic voting behaviors, one can now ask: how the situation of voting "ideologically" or "strategically" could affect the voting outcomes in FEFI? One can shed some light on this question by considering the hypothetical situation of putting the question of legalizing the gay-lesbian marriage to the public vote. This is indeed a highly sensitive issue to many voters and has great potential to maximize the instances of "irresponsible," "ideological," and "strategic" voting. Before presenting the example, a small note on characterizing strategic voting as "mischievous" voting behavior—as was done earlier in this section—is in order. When a voter votes strategically, he behaves "mischievously." That is to say, he does not vote for his true public wills, or according to his true wills as he is expected to do—(nor does he vote "irresponsibly" for the sake of his personal-interests, nor "ideologically" in accordance to his ideological or religious commitments)—but votes for a candidate or an issue which he opposes in order to eliminate or weaken another candidate or issue which he also opposes. Both the motive and the payoff in strategic voting are that it helps to strengthen the position of the voter's preferred candidate or issue in the next round of voting, or it renders some benefits to them indirectly. This pattern of voting is commonplace in the open presidential primaries in the United States.

There is, however, another pattern of voting that is essentially different from strategic voting, but could be mistaken for it. When there are more than two candidates or issues to choose from in a single-round (i.e., plurality) voting, a good percentage of voters vote for the "better" candidate or the better issue among the ones they do not support, when they are certain that their true political choice(s) cannot win. In these situations, voters do not vote for their true public preferences because they do not want their votes to "go to waste." Being certain that their preferred candidate would not win, they vote regardless, for they want to "exercise their voting power." This pattern of voting is often referred to as "voting for the lesser evil." Both the motive and payoff for engaging in this pattern of voting are to make the best of one's voting power, and at the same time, minimize the effects of perceived or expected evil. The lesser evil pattern of voting is usually an indication that the voters are not presented with a meaningful set and a wide range of options to choose from. In light of this distinction, one can argue that voting for the lesser evil cannot be regarded as mischievous voting despite the fact that the voter votes for something other than his true public or political choice. This pattern of voting is not mischievous mainly because the voter does not "conspire" to eliminate or weaken an undesirable candidate or issue in order to increase the chances of electibility for his preferred candidate or issue in the ensuing round, which is the case with strategic voting. Rather, he wants to make the best use of the voting power available to him in order to make the best *practical and realistic choice*. This pattern of voting, in order to distinguish it from strategic voting, will be referred to here as "realistic-practical voting."

It is important to emphasize that in "realistic-practical voting" patterns, the voter votes in accordance with a ranking order that he has devised for himself. The lesser evil he votes for is his second-, or third-, or fourth-ranked choice, so to speak, even if this choice stands far removed from his first choice, and could be next to the last choice available to him, i.e., the greatest evil itself. In strategic voting, on the other hand, one does not necessarily vote according to a ranking system, nor does he necessarily vote for the lesser evil (it might well turn out that the strategic voter would vote strategically for the greatest evil itself); but rather according to a strategy that is intended to help his issue or candidate of choice indirectly. Not only does the strategic voter "conspire" to eliminate a candidate or an issue by voting for his/her/its opponent, but he also uses the candidate he votes for as a means toward his end of improving the lot of his preferred candidate or issue.[21]

Now the hypothetical situation of putting the question of legalizing the gay-lesbian marriage to the public vote. Suppose the following are the policy options presented to voters:

A. Civil union of gay and lesbian couples should be allowed and be granted the status of "marriage" in the same sense as the marriage between a man and a woman. Such a union would be recognized legally for *all* purposes, in the same way the marriage of a heterosexual couple is—no exceptions are stipulated.

B. Civil union of gay and lesbian couples should be allowed but should *not* be granted the status of "marriage." However, such a union would be recognized legally for *all* purposes, in the same way the marriage of a heterosexual couple is—no exceptions are stipulated.

C. Civil union of gay and lesbian couples should be allowed but should not be granted the status of "marriage." However, such a union would be recognized legally for all purposes, in the same way the marriage of a heterosexual couple is; *except* that legally joined gay couples will not be allowed to adopt children.

D. Civil union of gay and lesbian couples should be allowed but should not be granted the status of "marriage." However, such a union would be recognized legally *mainly* for the purpose of holding joint property and inheritance rights (e.g., dividing up of joint ownership of property, house, etc., when the couple breaks up, as is the case in a heterosexual divorce).

E. Civil union of gay and lesbian couples should be allowed but should not be granted the status of "marriage." However it should be recognized legally *only* for the employment benefit purposes and similar matters (e.g., being a beneficiary in employee medical plans).

F. Civil union of gay and lesbian couples should not be allowed under any circumstances.

It is reasonable to assume that the overwhelming majority, if not all, of gays and lesbians, given that this is a vital and high-stake issue for them, would not comply with the prescribed procedures for voting responsibly, and instead would ("irresponsibly") place all of their 300 points on policy option A. Anticipating this voting behavior on the part of gays and lesbians, and also considering that this is an ideological issue for extremely religious members of the public (e.g., the Christian Coalition voting bloc), most, if not all, of those belonging to this bloc would not abide by the prescribed procedures either and would ("ideologically") put all of their 300 points on option F. However, the rest of the voters, one can argue, would follow the prescribed procedures in spreading their 300 points. Assuming that there will be a moderate to high voter-turnout (given that this is a controversial issue), and assuming that most voters are neither gays-lesbians nor extremely religious (and thus would not regard this as a high stake or ideological issue for themselves), one can argue that the real weight of the people's votes would be carried by those who would follow the proposed procedures and vote responsibly. In case no option gets to win the prescribed majority (e.g., 50%+1), the highest two vote receivers will go to a runoff vote.

In a low voter-turnout, especially if the opinions on the questions are divided, it could be the case that options A and F would be the two highest vote receivers and would face each other in a runoff. Regardless of which options would make it to the runoff stage, one would assume that in the runoff vote, gays and lesbians will vote for the policy option that would give them the most benefit and the Christian Coalition will vote for the least benefit option. In the runoff, again, the real weight of the votes would be carried by those who would follow the procedures, unless there is a low turnout. A low turnout, either in the primary or in the runoff vote, would indicate that the majority of the voters are not interested in the issue and thus would accept any outcome; in which case the resulting collective will would be formed, and rightly so, by those who see this as a vital issue for themselves (i.e., gays and lesbians on the one hand, and the Christian Coalition on the other hand). However, even if the voter turnout would be low, i.e., only a small portion of non-gays, non-lesbians, and non-extreme-religious groups participate in voting, given the reasonable assumption that the weight of the votes of gays-lesbians and Christian Coalition would be roughly equal, then again, one can argue that it would be the vote of "responsible" voters who would determine the outcome by tipping the balance in favor of one or the other.

As to the place of strategic voting in this example, given the wide range of alternatives presented to voters, one can argue that voting strategically would be quite an unreasonable and risky course of action in this scheme for either of the high-stake voting blocs. For instance, if most, if not all, gays and lesbians, and their staunch supporters predict that the general public will vote conservatively on the issue (and thus options A-D have a lesser likelihood of winning), and as a result, place their points heavily on option E in order to force a runoff vote, then they risk the prospect of not having any one of options A-D in the runoff vote. (Of course this is based on the assumption that the Christian Coalition would concentrate heavily on option F, which could send it to the runoff vote along with option E.) The same would be true if the Christian Coalition predicts that the general public will have a liberal attitude to the issue and advises its voters to vote heavily for option B strategically with the hope of preventing option A from going to the runoff stage.

"Practical-realistic voting," on the other hand, would not be as unreasonable or as risky as strategic voting but would still be quite risky. For instance, some gays and lesbians or their political allies, anticipating that the majority of voters will place most of their points on the median options (in this case options C-D), they might strategically divide up their 300 points between these options with the hope of having them sent to the runoff, thus eliminating option F. Indeed, this is risky, for it could backfire if the Christian Coalition members decide to vote ideologically and exclusively concentrate on option F, in which case the likelihood of the option F making it to the runoff would be much higher than that of option A. The situation can also be risky for the Christian Coalition if its members decide to place their points strategically on option E (just under the median), in case they are not sure whether option F would make it to the runoff stage, and if it happened that gays and lesbians voted by their true wills and placed all of their points on option A. Another consideration that makes strategic voting, or practical-realistic voting to a lesser extent, risky is that not all gays or lesbians, or Christian Coalition members, speak with the same voice on every issue; and thus it would be difficult to coordinate and execute strategic voting or practical-realistic voting strategies in a precise manner or on a grand scale in this voting scheme.[22] The political costs could be devastating if plans are not executed properly.

The example of gay-lesbian marriage—(indeed a controversial issue that can test citizens' moral commitment to voting "responsibly")—presented here was intended to make the point that, resorting to "strategic" voting, and to lesser degrees to "realistic-practical" voting, would not be a reasonable game plan in FEFI. This is due to the fact that the scheme presents to voters a meaningful and a wide range of policy options to choose from, as well as allowing them to express their wills in virtually infinitely many different ways. It is a contention of this book that these two components, in addition to having the power to integrate their wills directly into the decision-making process, would motivate voters to vote responsibly and continue to keep their faith and trust in the democratic system. Moreover, given the fact that there are virtually infinitely many different ways of voting, including an accountable and "official" channel for registering one's protest (the option of "protest vote"), the very idea of devising a voting strategy to beat the system would appear as unnecessary and undesirable to most voters. Finally, given the complexity of devising and successfully executing a workable and successful strategy, and considering the political risks and costs if the strategy fails, voting strategically as a bloc would not be an attractive choice to organizations or special interest groups either. These two considerations would discourage voters from straying away from the path of voting responsibly.

The Problem of the Manipulation of the Voting Agenda

Continuing with the hypothetical situation of putting the question of gay-lesbian marriage to public vote, it can be argued that the very structure of the options or the order in which they appear on the ballot in this example are manipulative in that they favor median options C-D, or at best, one of options B-E, and thus reduce the likelihood that the radical option A or the ultra-conservative option F would be the winner. This is indeed true, for the median options would seem to many voters as safer or better options. The order in which these policy options appear would manipulate the voter, and thus the outcome. An easy way to overcome this difficulty would be to use lottery to determine the order in which the policy options would be presented to voters. However, this randomizing of the order of options would not decrease the chances of the median options or increase the likelihood that option A or option F could be the winner. This is mainly because, as long as the prevailing "public political culture," or some segments or manifestations of it (e.g., the public's "psycho-cultural" attitudes and sentiments on controversial issues), views options A and F as non-median or outside of the mainstream (as is the case in present-day American society), these options will have lesser chances than options B-E.[23] What lies at the root of this difficulty for options A and F is that the prevailing "public political culture" does not always coincide with the constitutionally affirmed values and sentiments, in conformity with which, the agenda-setting and policy-formulation bodies are expected to draft the policy options.[24]

As to addressing the problem of the manipulation of the agenda and policy options by the agenda setters and within the agenda setting institutions, one can argue that the main component of this problem can effectively be minimized by instituting procedural-institutional arrangements that would assure voters would be presented with a *meaningful and complete list of policy options* to choose from on every voting occasion—as was the case in the example of gay-lesbian marriage. These institutions can be monitored by an effective system of checks and balances, as well as by keeping them responsive to direct feedback from voters. The "protest option" and its "veto power" would work effectively in this capacity. Another way of minimizing the manipulation of agendas would be to use a lottery system to select some, if not all, members of the *ad hoc* committees that would be commissioned with drafting agendas in the agenda setting institutions.[25]

However, the game of manipulation and strategic voting that could go on among the agenda setters themselves, and *within* the agenda setting bodies (e.g., in the committees), is a much harder problem to address. No matter how effective the system of checks and balances would be, one should be resigned to the fact that some degree of manipulation will always remain ineradicable in these types of institutions. Thus, one should only hope to minimize the problem to a sufferable level. Having made this point, one can further argue that "manipulation" also has a positive aspect to it and no agenda setting institution should be without it. That is to say, some form of influencing the issues and agendas is necessary and desirable in most agenda-setting situations in order to assure that issues are framed in accordance to, and in conformity with, the constitutional constraints and institutional-cultural values. Far from undermining democracy, this sort of "manipulation" would strengthen it, especially in situations in which the "public political culture"—which feeds, or ought to feed, on the constitutionally affirmed values and sentiments—is strained under the weight of tremendous events or controversial issues, and as a result, appears to verge on losing its composure.

Putting aside this positive or ideal aspect of manipulating the agenda, the negative side of the phenomenon becomes a real problem when the manipulation—and the strategic voting that accompanies it—becomes the means for pursuing sectarian and personal gains, or is used as a weapon in factional power struggles. This problem, if widespread, would make agenda setting and policy option formulation more a game of political skills, manipulation, and strategies than a true method of making a substantive democracy work. What is at stake here is the integrity of the process of agenda setting and policy option formulation as well as the integrity and trustworthiness of the individuals involved in the process. One solution to the problem, as formulated, would then be to argue that social-scientific knowledge, moral reasoning, and virtue ought to play an important role in agenda-setting and policy-formulation processes and bodies, to the same extent that political skills and legal knowledge play a part in the legislatures of present-day American liberal democracy. (This solution will be discussed thoroughly in Chapters 3 and 4.) The other solution, ideally in conjunction with the first, would be the vigilance of a sizable portion of citizens who participate in politics and keep an observant eye on the agenda setters and reject those whose performances fall short of affirming the integrity of the process, or genuinely reflecting the values of the institutions of democracy. Sending messages of disapproval to agenda setters by "protest option," or the "veto vote" mentioned earlier, would be one effective way of doing this. Another effective way would be to hold the agenda-setters responsible for justifying the policy options they propose to the public in public deliberation sessions.

The Problem of Revealing the Wills of the Voters "Truly"

It is now the time to discuss the potential objection that would question the "accuracy," "truth," or "correctness" of the results produced by FEFI. One would expect that this objection would be raised by "social choice," "rational choice," and "public choice" theorists who, in recent years, have succeeded in casting serious doubts on the usefulness of the institution of voting as a means of determining the will of the public, and through these doubts, have raised fundamental or "deep" questions about the whole enterprise of democracy in general, and the participatory conceptions of the enterprise in particular.[26] Given that the notions of the "*full expression*" of the voters' wills and their "*full integration*" into the collective choice-making process are among the essential components of the "conception of democracy as the political empowerment of the citizen," and also given that FEFI is the practical embodiment of these two notions, FEFI must be defended vigorously against this objection. The "theory of voting," a branch of the social choice theory, is replete with contentions and theorems that undermine the credibility of all voting outcomes and cast a dark shadow on the possibility of democracy itself.[27] According to the "theory of voting," all voting outcomes are "unstable" and "ambiguous," and thus "meaningless."[28] They are unstable because of what the "paradox of voting" reveals, according to which, the ordering of individual preferences cannot be translated "coherently" into the ordering of social preferences. Voting outcomes can also produce "Condorcet cycles"—which sometimes are referred to as the "indeterminacy problem."[29] The paradox of voting (or the "Condorcet paradox") finds its generalized statement in Kenneth Arrow's famous Theorem of General Impossibility that states, as William Riker puts it in an accessible language in his *Liberalism Against Populism*, "no method of amalgamating individual judgements can *simultaneously* satisfy some reasonable conditions of *fairness on the method* and a condition of *logicality on the result*."[30]

Furthermore, the voting outcomes are also ambiguous, as the contention continues, because different voting schemes can produce different outcomes for an identical set of voter preferences. Moreover, no voting scheme would produce a unique outcome from any given set of voter preference orderings unless it is "dictatorial or manipulative."[31] The other centerpiece of the enterprise of the theory of voting is Anthony Downs' contentions that, in a two-party system, "parties would rationally locate themselves at the center of the voter distribution"[32], i.e., "would adopt platforms near the median"[33]; and that "citizens typically have no interest in voting or in learning enough to vote in their interests even if they do vote."[34] Finally, to make matters much worse for the democratic theory, a second important corollary of the Arrow's Theorem, as stated by Iain McLean, claims that "[t]here can be no 'will of the people' in a multi-dimensional society. Whatever option the people choose, there is another which a majority of the people would rather have. This seems to rule out direct democracy in any large group of people; certainly in a nation-state."[35]

To defend FEFI against the theory of voting, the following strategy will be employed in the rest of this chapter. First, the remaining pages of this section will be devoted to responding to the main challenges raised by the theory of voting via responding directly to the contentions offered by William Riker in his *Liberalism Against Populism.* Riker makes the case for the theory of voting eloquently by stating the doubts and challenges posed by the theory in a bold and accessible language. As will be seen, in the course of defending FEFI against the potential objections of Riker, a more complex version of FEFI will be introduced. It will then be argued that this new version makes Riker's contentions irrelevant to FEFI. In the second part of this strategy, Arrow's Theorem will be discussed in some detail in a separate section and then, it will be argued that Arrow's Theorem should be ignored, for some of the premises and assumptions it requires are either too narrow, or lose sight of the moral dimensions of democracy, and thus cannot be imposed upon FEFI.

First, the case of Riker's *Liberalism Against Populism*. The fundamental contentions of this work can be summed up as follows.

1. The outcomes in any voting system depend on the voting method utilized. Given an identical profile of voting preferences, different voting schemes would produce different outcomes.[36]

2. Although it could be the case that a certain voting scheme in a given situation could work better than any other scheme (i.e., one method could produce a truer or fairer amalgamation of the voters' judgements than other methods), there is no general way of knowing which voting method would be the best for any given situation.[37]

3. Voting outcomes cannot truly represent a given profile of preferences also because of the presence, or rather omnipresence, of strategic voting.[38]

4. Alternatives presented to the voters are always manipulated by the agenda-setters. "[I]t is demonstrably possible for social choices to be determined not by the true values of voters but by the manipulation by agenda-makers."[39]

5. In light of claims 1-4 above, "we should never take the results of any method to be a . . . true amalgamation of voter's judgments."[40] Voting outcomes are "inadequate—indeed meaningless—interpretations of public opinion."[41] They "fail to make sense"[42], and are "hard to interpret."[43] "[W]e never know what an outcome means, whether it is a true expression of public opinion or not."[44] Moreover, the outcomes are also meaningless because "the majorities that make outcomes are themselves in flux."[45]

6. Also, in light of claims 1–4 above, "we should never take the results of any method always to be . . . fair."[46] Voting outcomes "fail to be fair."[47] The outcomes are unfair also because there are no fair ways of deciding which method to use. All methods are fair in that they "embody different ethical principles . . . [but] . . . there is no fundamental reason of prudence or morality for preferring the amalgamation produced by one method to the amalgamation produced by another."[48]

7. Thus, having shown that no voting scheme can discover or reveal people's wills, or amalgamate their preferences in a meaningful way, one then must declare democracy in its "populist" or Rousseauean interpretation—(i.e., "embodying the will of the people in the action of officials")[49]—as an "empty" notion and reject it.[50]

8. Finally, having demolished the populist-Rousseauean conception of democracy in the face of the inability of the voting enterprise to discover the public will, the only hope for democracy is the representative liberal democracy which uses voting to "control officials, *and not more*."[51]

It should be noted that the first three claims listed above are interrelated in that they all deal directly with *the actual voting process and its outcomes*. (The first two claims are sometimes combined under the rubric of the "multiplicity problem."[52]) The fourth claim, on the other hand, is independent of the first three claims and is concerned with a process that starts and ends *before* voting takes place. The fifth and sixth claims are the conclusion of the first four and raise serious questions as to whether voting schemes are worth using to begin with. Based on the fifth and sixth claims, the seventh and eighth claims reject what Riker calls "populist democracy," and affirm liberal democracy respectively. Now, since the questions of strategic voting and manipulation raised by the third and fourth claims were addressed somewhat adequately earlier in this chapter, they will not be discussed any further.[53] Instead, the rest of this section will be devoted to defending FEFI against the claims 1-2, 5-6, and 7-8. But before beginning to do so, one should emphasize that these claims raised by Riker only apply to situations in which there are multiple (more than two) alternatives to choose from—as is the case with the first round in FEFI. As to the situations with binary (exactly two) alternatives, there seems to be a consensus that the absolute majority method (50%+1) is a "fair" and unambiguous rule; provided, of course, that these situations occur, as Riker puts it, "naturally," i.e., without any institutional or social-procedural intervention to reduce the alternatives down to two.[54] However, Riker believes that this situation is only true "in the abstract" because the binary alternatives are "extremely rare" in reality.[55] The ambiguity or meaninglessness in question in voting situations in which there are multiple alternatives, arise, according to Riker, when the results are close, and different voting methods declare different alternatives as the winner. Unfairness in this situation arises when the alternative that ends up being declared the winner does not exhibit a clear and convincing superiority over the ones rejected, or when the alternatives are reduced down to two by human intervention. According to Riker, this reduction is "often (possibly always) in itself unfair."[56]

* * * * *

In defending FEFI against the difficulties posed by claims 1-2 and 5-6 above, one should start by reiterating what the scheme is designed to do. As was argued earlier, FEFI was intended to endow the voter with two sorts of powers: power to fully express herself and power to have her expressions fully integrated into the social decision-making process. The power to fully express oneself requires that the voter be given a meaningful and complete range of options to choose from—in which case she would be able to express her wills in virtually infinitely many ways. One consequence of this is that the voter would be able to express her views and wills with the degree of intensity she deems appropriate. On this ground alone, one can argue that FEFI does what it is designed to do, i.e., it gives to the voter the power of full expression. Moreover, FEFI integrates these expressions into the decision-making in the fullest possible way, by integrating every "spoken" point of the voter into the calculation of the total points for each policy option. The clear winner of the policy options (e.g., 50%+1), which could be produced either in a primary or in a runoff vote, would be declared to be the collective will produced on the issue put to public vote.

Now, to begin with, recalling from an earlier discussion, it was not a contention of this book that the collective will arrived at by FEFI would be the "correctly" or "accurately" revealed "will of the people"—as is the case with the notion of the "general will" in Rousseau's conception. Rather, the main intention of FEFI was to give a real and practical expression to the idea of the citizens' direct participation in collective decision-making. The fundamental *moral* idea that gave rise to FEFI did not conceptualize democracy as a system of governing in accordance with "the dominant wills" of the citizens, nor as institutional arrangements designed to discover these wills accurately. Rather, the primary purpose of the scheme was to allow individuals to experience political power directly, and do so on an equal basis. Thus, in light of this contention, one can further argue that the democratic merit and utility of FEFI would remain intact and untarnished if claims 1-2 and 5 are granted, i.e., if it can be shown that FEFI is an inadequate means for discovering "the dominant will" of the people. (As will be argued next, FEFI is quite adequate in this respect.) This is because the democratic legitimacy of FEFI does not rest primarily on the claim that the policy options it adopts necessarily represent the "dominant" views among the people—(i.e., the views that are coherent or strong enough to be regarded as "dominant" views might not exist among the people)—but on the contention that it enables individual citizens to speak their wills fully (and on an equal basis) and takes into consideration every spoken word of theirs in deciding the policy options. This is the primary (political) moral justification for FEFI and for the conception of democracy that gives rise to it.

As to addressing the concerns expressed by claims 5-6 (the meaningfulness of voting outcomes and the fairness of the voting process), an important assumption of the "conception of democracy as the political empowerment of the citizen" has, all along, been the existence of a socially and politically educated citizenry, a strong system of civic education, and an extensive set of public deliberation institutions. As will be discussed in Chapters 3 and 4, the latter could come about by instituting social-political arrangements that would give all of the contending views equal public voice and deliberative power before the actual voting takes place. The pre-voting deliberations would make the voting outcomes meaningful in that the outcomes would be the embodiment of, and the composition resulting from, the reflected-upon and reasoned judgements of a politically knowledgeable public. (One can further argue that the pre-voting deliberations would make the voting outcomes "epistemically correct."[57]) However, the winning policy option would not necessarily be the "true" representation of the dominant public will or the will of the majority—(granting Riker his claim that "majorities that make outcomes are . . . in flux")—but rather a revelation of the strongest or the most pronounced and shared expressions of individual wills *entered* into the scheme.

Now, in case it can be shown that FEFI is inadequate in revealing the most pronounced wills of the voters within some *reasonable margins of "accuracy"* or some *criteria of "acceptability,"* then one can argue that this problem would be *technical* in nature. And thus, it could be overcome by *technical solutions* and the appropriate sort of institutional-procedural arrangements that would *optimize* the output of FEFI in order to produce a "truer" or "fairer" approximation of the prevalent or broadly expressed wills of voters on the issue put to vote. One possible solution would be to use a different point system or to devise another collective will composition method that would perform more complex operations with points placed on policy options, rather than simply find their arithmetic sums, as is done by FEFI in its simple form presented earlier. It is also possible to use different amalgamation-composition algorithms for different voting situations or different issues.[58] A third possible solution would be to use a combination of different methods (e.g., Borda Count and Approval Voting methods in combination with the summation of points method used by the original version of FEFI) simultaneously in order to evaluate and compare with one another the total points accumulated by each policy option, and send to the runoff stage the two alternatives that would be ranked as the top two options by a specified number of these methods.[59]

Still a more complex solution is possible, which can produce a much "truer" and "fairer" representation of the wills prevalent among voters by giving them even fuller power in integrating their wills into the collective will-formation process. The solution in question, as will be seen below, would offer voters the opportunity to *vote interactively* by giving them the option to *revise* their votes twice. The method would offer voters the option of "practical-realistic voting" (if they choose to use it) in case their preferred policy options would end up being eliminated—doing so would prevent their expressed wills (or decision-making power) from going to "waste." What follows is a description of how such a scheme would work.

Suppose there are six policy options available—not counting the "protest option." This scheme would amalgamate-compose the input of voters in three rounds in the primary stage; and in each round, it would use a "combination algorithm" that would use different schemes simultaneously, as was suggested by the third proposed solution above. In the first round, the scheme would eliminate the two weakest options that end up being ranked last. In round two of the primary stage, the scheme would start from the beginning and would match the surviving four options against one another, using the same combination algorithm, and eliminate the last ranked (the weakest) option. (Here, one should note that, with only four options to choose from, the dreaded cycles become unlikely.[60]) Finally, in the third round of the primary stage, the scheme would compare the surviving three options against one another, again using the combination algorithm, and would declare as winners of the primary the options that end up being ranked at first and second places.

The scheme would work as follows. The voter, when voting, would distribute his 300 points among the policy options he prefers, and with the intensity he assigns to them, in three layers as prescribed by FEFI. However, before he actually drops his voting card in the ballot box (or rather before he pushes the button to electronically register his vote for vote counting purposes), he can use one or both of the following options available to him at that point. First, he has the option of having the points he is placing on the policy options, that could eventually end up being ranked fifth and sixth at the end of the first round of calculation, redirected and redistributed among the four options that would end up surviving the first round, in any way he deems appropriate. His second option would be to have the points he is conditionally redistributing among the four would-be surviving options at the end of round one redirected again; that is to say, to have the points he has conditionally placed on the last ranked alternative at the end of round two be redirected toward the three options that would survive it in any way he prefers.[61]

The top two point receivers at the end of the third round in the primary stage would be declared as the winners of the primary and would be sent to the runoff stage. Clearly, there would be a need for a powerful and fast computer algorithm that would do these calculations and would receive the instructions from voters on redistributing their points in the second and third rounds *at the time of voting*, and *before* it starts the calculations for round one and the subsequent rounds. Once round three is completed, the "democracy computer" would report the outcomes of the calculations for each round and declare the two policy options that have survived this process of elimination as the winners of the primary—which would then face each other in the ensuing runoff stage.

Admittedly, the process of giving to the "democracy computer," in advance, the instructions for the redistribution of points in rounds two and three can be quite complicated and would involve filling out a long computer questionnaire that could logically challenge voters with a long series of fill-in-the-blanks statements such as: "If the policy **Option I** which I am placing **55 Points** on does not survive the first round, then redirect these points to policy **Option V** if it survives; otherwise place them on policy **Option III**." This complication can be overcome easily by programming the "democracy computer" to return to each voter at the end of each round of calculation and ask for instructions on redistributing those points of his that could go to waste. The following is an example of how the "democracy computer" would return to each voter via e-mail and ask for instructions for round two:

> Dear Nellie Conrad,
>
> I completed the first round of the amalgamation for the vote on the "Budget Plan Issue (2036)" at 3:23 am on 7/12/2035. The policy **Option VI** that you had placed **90** points on, as well as the policy **Option II** that you had devoted **55** points to, were eliminated in the first round. I am sorry, I cannot tell you how the remaining four policy options are ranked for fear that it might influence your decision. Please fill out this questionnaire in order to instruct me as to how I should redistribute these **145** points among the surviving policy **Options I, III, IV, and V** by the deadline of 11:59 pm on 7/13/2035. I will start the calculations for round two at 2:00 am on 7/14/2035. If you do not respond by the specified deadline, you will lose these points.

At the end of the second round, the "democracy computer" can return to the voter with a similar message asking for instructions for the third round. Although having the "democracy computer" take the voting input in bits and pieces makes voting less complicated, it makes the decision-making process take a week to complete. Despite this undesirable aspect of this last solution, its advantage over the third one suggested above is that it executes the elimination process in three successive rounds rather than eliminating four options at one swift calculation. It first eliminates the two options that end up being ranked last in the "combination algorithm," and thus minimizes the "unfairness" and "untruth" that can result from eliminating the third- and fourth-ranked options very early on, especially when the first and second ranked options do not beat them in all of the individual schemes used in the combination algorithm. The same is true in rounds two and three.

This fourth solution would largely eliminate the possibilities of strategic or practical-realistic voting in its initial round. Voters would vote according to their true wills initially in the first round, for they are assured that the points they place on their first choice option(s) would not be wasted, in case these options are eliminated in the first round. If this turns out to be the case, voters can revert to practical-realistic voting in the second and third rounds, if they choose to do so—which could be an appealing option to many voters. Furthermore, given that the actual ranking of alternatives will not be disclosed until the process is completed, strategic voting would not be a reasonable option. Moreover, one can argue that reverting to, and engaging in, realistic-practical voting in the second and third rounds can be conceptualized as a process through which *voters would engage in a sort of indirect negotiation among themselves* in order to arrive at two "compromise wills" by channeling their wills into two policy options that would eventually be sent to the runoff stage when there is no consensus nor a clear "dominant will" among them. Finally, having conceptualized this more complex version of FEFI as a vehicle of indirect negotiations for reaching two "compromise wills"—and granting to Riker his ontological skepticism about the existence of "dominant wills"—one now can argue that FEFI *should be viewed not as a "revealer" or "discoverer" of the "true" and "dominant" will of the people, but rather as an amalgamator-composer of such a will in a democratic fashion.*[62]

These solutions are only four among numerous ways of addressing the difficulties associated with the enterprise of voting that claims 1-2 and 5-6 bring to light. With a little imagination, one can devise other methods that can minimize these difficulties further. It should be noted that these solutions, as well as the difficulties they are designed to address, *only* apply to the primary stage of the proposed voting scheme, in which voters are given multiple policy options—(e.g., the gay-lesbian marriage example discussed earlier)—and *when* no policy option clearly beats all others according to some prescribed criterion (e.g., the absolute majority rule).

To recap, the arguments above that gave rise to these solutions rested on the assertion that claims 1-2 and 5-6 do *not* pose *moral* challenges to FEFI, *but only social-technical* ones—i.e., institutional-procedural, as well as social-scientific and mathematical ones. It was argued that the "conception of democracy as the political empowerment of the citizen" is not threatened by these claims at all; for the conception rests on the *moral principle of the direct, full, and equal participation of all*, and *not* on the claim of being an accurate revealer of the dominant public wills. Having restated the argument, one should add that the "conception of democracy as the political empowerment of the citizen" *does rest on an implicit assumption that an "adequately democratic" scheme of collective decision-making is possible*; that one can attain a "reasonably fair and accurate" method of amalgamating-composing the collective wills out of the expressed wills of the citizens.

In addition to the arguments presented above, one can further argue that the question at hand is not exclusively about the *reality* of the fairness or accuracy of the voting methods and outcomes, but also about the *public consensus*, and the *public justifications* for them. That is to say, it is also the question of whether the majority (hopefully the overwhelming majority) of voters believe that the voting scheme in use is "reasonably fair and accurate," or satisfies some acceptable criteria of fairness and accuracy, and that there are good reasons for choosing it over other methods. No doubt, high levels of the fairness, accuracy, truth, and the consistency of the outcomes are highly desirable in every voting situation, but not absolutely necessary. That is to say, *a moderately accurate scheme that can be reasonably justified to a politically knowledgeable and participatory citizenry, and is regarded by it as acceptable, would be sufficient.* It should be pointed out that the present-day American liberal-democracy rests on the same implicit assumptions about the voting process and voting outcomes. The final result of the U.S. presidential elections in the year 2000 is a good case in point. Putting aside the controversy over the accuracy of vote-counting in Florida, the relevance of this example lies in the fact that the public political culture of the United States accepted the election of George W. Bush as the president of the United States, even though he did not represent the "dominant will" of the public (he won less than 50% of the votes and about a quarter of a million votes less than Al Gore). Bush won according to the election rules which required the winner to win only the Electoral College and not the popular vote. Once it was done and over with, numerous polls showed that the majority of the population believed that the process, despite the controversies, was fair; for it followed the established procedure of American democracy for electing presidents, which the public political culture does not regard as being blatantly unfair or unacceptable.[63]

In summary, the force of the arguments presented above was directed toward weakening the appeal of claims 1-2 and 5-6—and to some extent claim 3—that *all* voting schemes are unfair or that they *all* produce unfair or untrue outcomes.[64] These arguments attempted to show that some methods are "fairer" than others and produce "truer" outcomes. But more importantly, these arguments were intended to make the case that the issues of fairness and accuracy in the voting scheme proposed earlier (FEFI) are to be treated as quantitative concepts, in just the same way as the conception of democracy as the political empowerment of the citizen is conceived. Putting it differently, *the fairness or accuracy of the outcomes in* FEFI *is a matter of degree, and not an attribute or quality that an outcome can either lack or possess*. Moreover, since there are no qualitative solutions to the problem (granting the theory of voting its landmark achievement), one, therefore, should place his hope in quantitative improvements. Furthermore, granting to the theory of voting the claims that unfairness in the process is ineradicable, and completely accurate determination of the public wills is unattainable (the so-called indeterminacy problem), it was argued that one should then hope to mitigate the difficulties claims 1-2 and 5-6 allude to by *solutions that are technical in nature.*

(As will be seen in the discussion of the Arrow's Theorem in the following section, FEFI does not encounter the "indeterminacy problem.") By finding appropriate technical solutions to these problems (i.e., by finding appropriate institutional-procedural arrangements, social-scientific methods, as well as the right mathematical relations to be used in amalgamating-composing the inputs of citizens), one can minimize the unfairness involved in the process, and at the same time optimize the truth of what the outcomes reveal. Finally, these arguments, and the solutions they produced, were premised on the contentions offered in two previous sections; viz., the difficulties illuminated by claims 3-4 (manipulation of agenda and strategic voting) are minimizable and can be kept in check.

* * * * *

As to the seventh claim, what Riker defines as "populism," and rejects as being "empty," does not have much in common with the conception of democracy as the political empowerment of the citizen; and for this reason, it need not be confronted. What Riker takes as populism, it suffices to say, is a representational-constitutional system, which also is the case in liberal democracy. The difference between the two lies in that, in Riker's populist regime, an individual, or a group of individuals, lays claim to being the embodiment of the popular will, and begins to tamper with the constitution or suspend it in order to make way for what Riker calls a "speedy embodiment" of that will. More than any other regime, his description of populism seems to fit the profile of Argentina under Peron and Peru under Fujimori.[65] Moreover, it must be commented that Riker's characterization of this conception of non-participatory "populism" as Rousseauean or as populism is highly problematic and draws sharp criticisms.[66]

Finally, the eighth claim. Riker arrives at his conclusion—that liberal democracy is the best possible form of democracy—by a process of elimination. The problem with the way Riker employs this process lies in that he does not begin with many choices. It is either the "constitutional dictatorship" of populism *or* the constitutional democracy of liberal democracy.[67] In this contest, no doubt, the latter would win hands down. Riker's line of reasoning in endorsing liberal democracy is very easy to follow. In light of the problems he illuminates about voting schemes, he is inclined to expect the least from them. As a result, he opts for a *minimal* conception of democracy that expects a minimal level of participation from citizens: all citizens have to do in the way of participation is "to control officials, *and no more*."[68] In this, Riker, and social choice theorists in general, are quite consistent. To grasp this consistency fully, one needs to bear in mind that these theories operate within the limited political scope of liberal democracy and its patterns of representation and electoral arrangements, as well as working with self-interested conceptions of the individual that dominate the theoretical-conceptual frameworks of both liberal democracy and the discipline of economics. A good case in point would be Anthony Downs' results that are premised on taking voters and the political candidates as primarily self-serving, and rely on the existing inadequate methods of expressing the voters' wills.

Arrow's Impossibility Theorem

Before closing this chapter, one needs to address the "challenge" of Arrow's Impossibility Theorem, and explain how FEFI can go around the potential difficulty posed by it. In what follows, this challenge will be met in two ways. First, it will be argued that Arrow's Theorem should not be regarded as a *real* challenge facing the democratic theory in general, and the democratic utility of FEFI in particular. This will be done by calling into question the plausibility of some of the fundamental premises and methodological assumptions of the "theory of voting." Second, it will be argued that Arrow's Theorem itself is open to questions insofar as some of its own assumptions and conditions are concerned. These questions, as will be argued, weaken the appeal of the theorem considerably and show that it has a narrow scope. Taking these arguments as a backdrop, it will then be argued that Arrow's Theorem should not be promulgated, as is widely done, as a theoretical "barrier" that precludes the possibility of devising democratic schemes for collective decision-making, including FEFI. Before starting this task, it should be mentioned at the outset that throughout this section, the "theory of voting" will be treated as a branch of the social choice theory, and at some instances, will also be used interchangeably with the theory itself.[69]

The true merit of the theory of voting and its landmark achievement, Arrow's Impossibility Theorem, lies in that they bring to light the problem of interpreting the meanings of voting outcomes, which for the longest time went unnoticed. One positive outcome of having the difficulties that Arrow's Theorem illuminates (e.g., the possibility of cycles) out in the open has been to compel the theorists of democracy to seek higher standards of rigor in their works. This merit aside, one should point out that the theory of voting and Arrow's Theorem themselves suffer from a host of theoretical and methodological difficulties that one needs to bring to light. These difficulties, as will be seen below, undermine the intellectual-theoretical credibility of the enterprise of the theory of voting, on the one hand, and raise questions about the real-life applicability of the results it produces, on the other hand.

To begin with, a main problem with the theory of voting has to do with the overall theoretical-conceptual framework within which it operates. The theory of voting can be characterized negatively as the disease-ridden child of the marriage of political science and welfare economics, in whose ill-formed mind the *political-philosophical* question of *democratic* decision-making is contorted into the *political-economic* question of aggregating the *rational preferences* (i.e., *private* interests) of purely self-interested individuals into a single *social choice*. (The social choice produced thus would be the aggregation of the private wills.) The question of democracy in this enterprise has become the object of study by a discipline whose theoretical and conceptual tools are primarily developed for the purpose of approaching the world from the standpoints of cost-benefit calculations, equilibrium analyses, and rational game strategies. Thus, the fundamental problem that besets the project of the theory of voting is that, within its theoretical framework, the burden of addressing the *moral* question of *democratic* decision-making has been processed into the *logical* task of performing gain-loss analyses and devising rational (i.e., logical) strategies of decision-making or social organization.

This larger meta-theoretical problem aside, one should now point to the problematic nature of some of the fundamental assumptions and methodological tools that the theory works with. As to its underlying assumptions, the theory of voting starts with the premise that there exists a somewhat exact analogy between the political system, on the one hand, and the economic system, on the other hand. This is evident in postulating voters as self-interested political-consumers, and politicians as political-profit-seeking entrepreneurs. A similar premise underlies the theory's imposition of free-market conditions on the political system. The theory assumes that there exists a free and competitive political market wherein the political entrepreneurs compete for the votes of political consumers. One main difficulty with this analogy is that, as numerous commentators have argued, it is highly implausible.[70] Another related area of difficulty has to do with the contention that this analogy works to "debase" the very idea of democracy and "invite[s] its subversion."[71]

Some of the difficulties that arise from applying economic concepts to politics are discussed by Emily Hauptmann in her *Putting Choice Before Democracy*. As Hauptmann argues, the "rational" in "rational choice" is understood as instrumental rationality, i.e., as the capacity of the self-interested individual to adopt means to achieving her *private* ends.[72] Applying this notion of rationality to politics, especially to voting, becomes problematic; for the notion does not, and cannot, address the question of social commitments and public responsibilities.[73] A good example here would be the case of those voters who vote, not instrumentally (e.g., hoping to gain benefits, draw "psychic gratification," or contribute to the sustenance of democracy), but "out of a sense of civic duty."[74] Another example would be the notion of "choice" and its identification with "preference" in the social choice theory.[75] Political choices, or political wills expressed by voters, one can argue, should not be regarded as "rational choices" or "rational preferences"; for political choices, unlike economic preferences, cannot be modeled as necessarily rational. The former are more related to "commitments" and not strictly to personal utilities, as is the case with the latter.[76] As a consumer, the individual chooses among the competing goods and services according to some preference profile concerning the economic criteria of cost, durability, etc. A problem arises when one applies this understanding of choice to politics. As Hauptmann points out, "[t]here is something odd about speaking of choosing our commitments in the sense in which rational choice theorists would have us understand choosing."[77] A second problem with the notion of choice in the theory of voting has to do with the fact that most choices made by individuals in choosing economic goods and services only affect the individuals themselves; this is quite unlike making political choices, e.g., voting, when the individuals' choices "have a much more direct effect on others than on themselves."[78]

As to the methodological difficulties, the theory of voting relies on some unrealistic assumptions about the ways real individuals approach the question of choosing among competing alternatives in voting situations. The "transitivity property of preferences" is the most important of these methodological assumptions. In fact, this property is the main "scientific deductive" tool used in deriving Arrow's Impossibility Theorem, the most important result of the theory of voting.[79] According to this property, if a voter prefers option A to option B and B to option C, then she must also prefer A to C. Against the practice of using this property with the theory of voting, one should argue that the ordering of social-political preferences in *real*-life voting situations, rather than being dictated by this property, are conditioned by the *contexts* of options presented to voters. That is to say, if voter x prefers A to B, B to C, C to D, and D to E, and thus—by the transitivity property—C to E, one should not deduce logically that she would still have preferred C to E, had option D not been on the ballot. This is because the absence of D could have led the voter x to choose a different ordering of preferences in which case it could have been possible that she would have preferred E to C. Far from being conditioned by any social-political or social-psychological constraints or contexts, the transitivity property of preferences in the theory of voting is assumed to be applicable with the same degree of rigor it is used in mathematics.[80] And here lies the problem with the transitivity property of preferences: how can one take a purely mathematical property as governing the voters' political choices?[81]

The response offered by the theory of voting to this criticism is that voters are postulated as "rational" choosers. As McLean puts it: "[a] minimal definition of 'rationality' is 'transitivity of preference[s]'."[82] Notwithstanding the reasonableness of assuming that most voters would find the transitivity property reasonable, there are no good reasons to believe that they would, or do, use it in ordering their political choices. And if they are inclined to use it, there are no reasons to believe that they apply the property with the same mathematical rigor as the theory of voting itself does. Clearly, individual political choices are not abstract entities; rather, they are the products of complex processes that often defy mathematical reductionist formalizations.

In addition to being ill-suited for modeling the real-life situations, the transitivity requirement also creates a theoretical difficulty for the theory of voting. There are good reasons for doubting that "rational" choice-makers must necessarily apply the property of transitivity to the ordering of their choices in each and every case. In some cases, the situation could be such that rational choice-makers might not be able to apply the property, for rationality or reason would dictate that other evaluative criteria be applied instead. Identifying rationality primarily with transitivity, or requiring that a rational choice-maker's ordering of choices be transitive, forces the theory of voting—(and Arrow's Theorem as will be seen shortly)—to embrace utilitarianism as its moral theory on an issue where utilitarianism is most vulnerable and least tenable, viz., the issue of dealing with values.[83] That is to say, if a rational choice-maker is obligated to apply transitivity in her ordering of choices, because there are "*logically* good reasons" for doing so (primarily, but not exclusively, the rationality of maximizing her utilities), this would imply that she might have to trample upon some "*morally* good reasons," e.g., political-moral commitments, that would guide her to use some evaluative criteria other than transitivity.[84]

Finally, the other shortcoming of the transitivity property, as used by the theory of voting, is that it only allows an ordinal ordering of choices, thus excluding cardinal ordering or any other form of ranking that accounts for the "weights" or "strengths" of choices. That is to say, unlike its counterpart in mathematics, a transitively ordered relation among a set of alternatives cannot provide any information as to the distances between different alternatives. This would mean that the theory of voting does not incorporate the intensity of the voters' choices into its formalization.[85] (This is also the case with Arrow's Impossibility Theorem.)

The mathematical formulations of the voters' behaviors, with which the theory of voting works, would not be possible without employing these questionable assumptions and methodology. The same can be said about the paradoxes the theory generates. These paradoxes arise in part as the result of the usage the theory makes of these problematic assumptions and methodology in its quest to frame the question of democracy as a mathematical-economic problem. These paradoxes result also in part from posing questions, as Benjamin Barber puts it, "in the vacuum of abstract rationality, where they are stripped of historical and political context and removed from the arena of will and political judgement."[86]

In light of these considerations, one is justified in claiming that, more than anything, the paradoxes produced by the theory of voting bring to light the paradoxical nature of the theory itself, as well as the paradoxical nature of the purposes it intends to serve. As Emily Hauptmann incisively notes,

> [o]n the one hand, rational choice theorists identify democracy with honoring individual choices, a norm they believe has been overshadowed by pursuing what to their minds are the dubious goals of securing the common good or increasing popular participation. On the other hand, they also conclude that choices citizens are given are not worth making because they are either too insignificant individually to make any difference or are offered and counted in ways that end up distorting the very things that were supposed to be honored. The second position ultimately undermines the first.[87]

Now, putting aside the problems associated with the questionable methodology and assumptions, and also ignoring the paradoxes that haunt the theory of voting, one can raise another question as to the real-life applicability of the results the theory produces. Not only are the assumptions the theory works with unrealistic, they are also constrained by some conditions and limits that remove the theory further from reality. A good example of such limits is that, as will be seen below, the theory of voting works with situations in which there are only a few voters and a few alternatives. Given these limitations, and given the theory's highly abstract nature, the paradoxes produced by the theory seldom materialize in real-life.[88] Finally, as to the question of limits, John Burnheim, speaking within the context of discussing Arrow's Impossibility Theorem, makes the following salient observation about the theory of voting:

> It is very much less clear how relevant this knowledge is in a practical context. On the one hand, most social decision-making is at a level so far from theoretical limits that those limits may well be as irrelevant to our purposes as absolute limits on the velocity of matter are to designers of motor cars. On the other hand, the most important practical difficulties are precisely those that are brushed aside in the theory, which tends to assume such conditions as perfect knowledge and rationality.[89]

* * * * *

Having raised some questions regarding the overall philosophical-theoretical soundness of the enterprise of the theory of voting, as well as raising some questions as to the real-life applicability or relevance of the results it produces, the attention of the rest of this section will be focused on Arrow's Impossibility Theorem, which is often regarded as the landmark result produced by the theory of voting. To begin with, it should be stated that the theorem is valid only if one admits its assumptions, methodological tools, and restrictive conditions. Now, as far as its assumptions and methodology are concerned, Arrow's Impossibility Theorem works with pretty much the same set of rules and assumptions that the theory of voting and its "mother" theories (i.e., the rational choice, social choice, and public choice theories) employ. As to the conditions, the theorem requires five. These conditions are often divided into two groups, one being the "conditions of fairness," and the other, the "condition of logicality."[90] What follows are the statements of these conditions:

There are four conditions of fairness:

1. *Universal domain* (U)—According to this condition, if there are *n* choices available and there are *m* possible ways of ordering them, then the voters should be allowed to choose any one of these orderings. For instance, if there are three choices A, B, C, then there are thirteen possible ways of ordering A, B, C; and voters should be allowed to choose any of these thirteen possible ways—in other words, the domain contains thirteen possible orderings and voters are allowed to choose any one of them.[91]

2. *The weak Pareto condition* (P)—This condition requires that if every voter ranks some option C over some option B, then the social ordering should do the same. In other words, social ordering should not rank B before C.

3. *Independence of irrelevant alternatives* (I)—According to this condition, the social ordering of any pair of two options, e.g., A and B, should depend only on the individual orderings of A and B and should not be affected by (or should be "independent" of) the individual orderings of either one of A or B versus some

other ('irrelevant") option(s), such as C, in the domain of options. Alternatively stated, A and B should only be compared pairwise, and *without* considering the information on how A or B would compare in pairwise comparisons with other options in the domain.

4. *Non-dictatorship* (D)—This condition implies that if one voter ranks some option C over some option A, and every other voter ranks A over C, then social ordering should not rank C over A.[92]

There is only one condition of logicality.

> *Collective (or Social) Rationality*—(CR). According to this condition, *social* ordering must be both *transitive* and *complete.* (This is in addition to the requirement that *individual* orderings must be both transitive and complete.)

An ordering (be it individual or social) is said to be complete "if it can say of every possible pair of options whether that one is better than the other or that they are equally good."[93] For example, if there are four available options A, B, C, and D, a complete social ordering would indicate how each option, say A, stands to each of the other options in pairwise comparisons. For instance, voter x prefers A to B, but not to C (which means he prefers C to A). Moreover, he is indifferent when it comes to comparing A with D. Given the requirement that orderings must also be transitive, the ordering profile for voter x in this example would be: C first, either of A or D next (A=D), and B last. Moreover, given the requirement of completeness, the following statements, which are consistent with the transitivity requirement, must also be made: B is not preferred to A, or to C, or to D; C is preferred to each of A, B, and, D; there is no difference between D and A; C is preferred to D; and D is preferred to B.

Having provided a description of each of its conditions, Arrow's Impossibility Theorem can now be stated as follows:

> If there are at least three voters and at least three options, there exists no procedure (i.e., no welfare function) for deriving a social ordering from individual ordering that can simultaneously satisfy all of U, P, I, D, and CR.[94]

It is of utmost importance to note that the theorem does *not* state that its conditions cannot be satisfied simultaneously for *any* given profile of voters' preference orderings, but that it cannot guarantee the simultaneous satisfaction of all of the conditions for *all* possible profiles of the voters' preference orderings—(i.e., for all possible profiles of the voters' preference orderings *theoretically* allowed by condition (U), which might or might not be materialized in practice). In other words, the theorem states that for any given voting occasion with at least three voters and three options, there exists the *logical possibility* that the specified conditions may not be simultaneously satisfied, and that social ordering could be intransitive or cyclical (thus producing an indeterminable outcome, the "indeterminacy problem"), and *not* that the ordering will necessarily be cyclical, or that it will have a significantly high empirical probability of being cyclical.[95] In this sense, Arrow's Theorem should be regarded as only a theorem about the *possibility* of cycles or indeterminacy, and *not* about the absolute impossibility of simultaneously satisfying Arrow's criteria of fairness and reasonableness—and thus *not* necessarily about the impossibility of democracy as it is often portrayed as "proving."[96]

As Riker himself notes, there are two ways of challenging the theorem. One is to inquire into whether the theorem has "practical importance."[97] The other is to challenge the theorem theoretically by calling into question the reasonableness of its conditions. In order to argue that the theorem has no *practical* importance, i.e., to show that it is not applicable to situations with a large number of voters, one needs to show that the empirical probability of encountering cyclical or intransitive orderings in large elections is insignificant. In an attempt to close this route of challenging the theorem, Riker puts forth a pre-emptive argument to the effect that cycles are commonplace.[98] In advancing his argument, Riker makes a number of simplifying assumptions. However, he admits that doing so "limits [the scope of] the interpretation severely."[99] A main difficulty that has haunted Arrow's Theorem since its inception is that its proponents fail to demonstrate, on realistic grounds, that cycles are of significant empirical probability for large elections.[100] In short, the claim that cycles are commonplace in large elections has not been substantiated. To the contrary, there have been a number of claims in recent years to the effect that there now exist ample "computational evidence," "simulations," and "empirical studies" to demonstrate that cycles in large elections "can hardly occur" or that they "are of no practical importance."[101] For instance, Gerry Mackie claims that,

> every developed and published example of a political cycle has now been challenged . . . after fifty years of scholarship, from the first publication of Arrow's theorem, no one has satisfactorily demonstrated the existence of a normatively troubling cycle in the real world.[102]

As to challenging Arrow's Theorem *theoretically*, one common route has been to challenge the conditions (CR), (I), and (U) of the theory.[103] The main way to challenge (CR) is to question whether it is reasonable to require that social orderings be transitive. That is to say, to argue that "Arrow's Theorem is about social *orderings*, [and] not [about] social *choices*." In other words, to argue that what matters in the real-life voting situations is that we only need to declare a *choice* as the winner (as the collectively decided decision), and not to produce an ordering of choices.[104] Stated alternately, in real-life voting situations, what matters is "the selection of a single alternative, [and] *not* the ranking of all alternatives."[105] Moreover, it has been argued that beneath the condition (CR) lurks some "fundamental philosophical issues" that need to be addressed. The main difficulty here is that the very idea of "collective (or social) rationality"—undergirding the condition (CR)—itself is problematic, for it implies the odd assumption that a collectivity (or society) can be treated as an individual or as having "an organic existence apart from that of its individual components."[106]

The other common way of challenging Arrow's Theorem theoretically has been to question the reasonableness of condition (I). One main problem with condition (I) is that, as it is argued by some, it "forces voters to assume *a priori* that there are no relations among the alternatives . . . which they might consider relevant in their choice [of one alternative over another] . . . this is for the voter to decide *a posteriori.*"[107] Another main difficulty, which seems more relevant to the discussion of FEFI, is that condition (I) shrinks the question of fairness in making social choices (and via that the larger question of democracy) to that of the *ordering* of alternatives via *pairwise comparisons.* In so doing, condition (I) brushes aside as deficient in fairness any scheme that uses an evaluative criterion other than the pairwise comparison of alternatives. Still a third difficulty with the condition (I) is that it disallows any scheme that permits value or intensity assignments to individual choices, or lets voters express their "strengths of preference."[108] Among the schemes excluded are two important methods of collective decision-making, viz., the Borda Count, and what is known as the "summation of cardinal utilities method." The first works with ordinal rankings of individual choices, the other with the cardinal rankings of these choices.[109] Via use of an example, Kenneth Arrow attempted to bolster the "reasonableness" of condition (I) by showing how unreasonable the Borda Count method could be.[110] The conclusion that Arrow drew from the example has been challenged since then.[111] Moreover, it has been argued that the Borda Count method is the "optimal" method, and it is "unique" in that it can "represent the true wishes of the voters."[112] It has also been argued that the Borda Count method renders the same decision outcomes as the pairwise comparison of alternatives (Condorcet method) required by condition (I).[113]

Finally, in challenging the condition (U), some have questioned the need for requiring an "unlimited scope," as specified by the condition, and argued that "not every conceivable combination of individual preferences need be considered in devising a social decision procedure, since only some would come up in practice."[114] Thus, allowing some reasonable types of "pattern restrictions," it has been argued, can make it possible to find "some ways out" of the difficulty posed by Arrow's Theorem.[115] These pattern restrictions (e.g., "single-peakedness" and "value-restriction") can make it possible to produce coherent social choices, provided that, as Trachtenberg has stated, "the members of society's preferences are not utterly disparate, but conform to an appropriate pattern."[116] That is to say, if there exists a "shared . . . understanding of how to compare alternatives one with another," and *not* necessarily the requirement that the preferences in question be widely shared among the members.[117]

In light of what has been presented thus far in this section, it becomes apparent that Arrow's Impossibility Theorem should be treated as a theorem *only* about the *logical possibility* of encountering difficulties in producing *transitive* social *orderings* of *preferences* from the voters' individual orderings—(and one should emphasize, within the confines of the conditions the theorem establishes)—and not about the high empirical probability of running into intransitivity in these *orderings*. Nor is it about the impossibility of devising *reasonably* "fair" and "accurate" methods of social *decision-making*. Amartya K. Sen puts the case elequently:

> In examining social decision mechanism, we have to take the Arrow conditions seriously, but not as inescapable commandments. . . . The issue is not the likely absence of rationally defendable procedures for social decisions, but the relative importance of disparate considerations that pull us in different directions in evaluating diverse procedures.[118]

One can push the argument a step further. Arrow's Impossiblity Theorem should not be interpreted as a proof of the impossiblity of democracy as Riker and others have attempted to portray, but rather as a proof that the question of democracy cannot be approached through the narrow scope of a preference-based and consequentialist logic that takes individual utilities (welfare) as the only object of value.[119] But questions still persist: why does the Rikerite type of interpretations of Arrow's Theorem arise? And why do they gain intellectual respect and credibility? The answer to the first question should be sought, in part, in the failure to grasp the *moral* substance of the idea of democracy, and in part, in the desire to reduce the complexities of the social world to formal and pseudo-logical categories of mathematical economics. As to the second question, the answer lies not just in the force of the arguments or the rigor of the axiomatic-analytic methods these interpretations employ, nor just in the fascination of some minds with the mathematical pyrotechniques they display in modeling social-political phenomena as mathematical-economic objects, but primarily in the ubiquity and appeal of the essentially economic (and market-based) conceptions of the social world, and life in general, *and* in part, in the appeal of a certain *value* system that undergirds the standards of rationality in the social choice theory.[120] One can go a step further and argue that the same types of reasons can be utilized to explain the rise of Arrow's Impossiblity Theorem in the first place. In their discussion of Arrow's Theorem, Friedland and Cimbala formulate a somewhat similar judgement:

> Arrow's proof is actually a demonstration of the logical inconsistency among a set of *value* postulates and *not* of some inconsistency between the rules of logic and non-dictatorial [i.e., democratic] methods of reaching social decisions. The assertion that social orderings ought to be transitive [condition (CR)] or that they ought to be independent of irrelevant alternatives [condition (I)] are hardly principles of logic. Although they coincide with notions of rationality often ascribed to, they are no less value judgements than the statements that a dictatorship is undesirable.[121]

*　　　*　　　*　　　*　　　*

Now returning to the question posed in the beginning of this section, viz., how FEFI could go around the "barrier" of Arrow's Impossibility Theorem, the answer would be that FEFI can ignore the theorem for two reasons. First, FEFI does not accept, nor does it comply with condition (CR), for (CR) is only relevant if one agrees to the applicability of the transitivity property to the ordering of individual preferences. The framework within which FEFI is devised, and the purpose for which it is intended, is radically different from the problem that Arrow's Impossibility Theorem addresses. It then follows that neither the transitivity of individual preferences nor condition (CR) is relevant to FEFI. The reason for this assertion is that FEFI is not an "ordering" scheme, let alone a transitively based one. The voter does not enter a *transitive* and *complete* profile of "preferences" into FEFI. Rather, she spreads her points among the options in any way she judges as proper. It might well be the case that she spreads her points only on one or two options and ignores the rest. Therefore, in this sense, her "preference ordering" (if one grants the applicability of these terms to FEFI for the sake of argument) is not complete. Moreover, given that the voter in this scheme "orders" her "preferences" in three different layers—(and as a direct result of this, these orderings can be different in each layer and even could contradict one another)—her overall voting profile on any given issue may have internal cycles and inconsistencies, and thus cannot necessarily be transitive either.

In making this argument for FEFI, one should add that these internal cycles or "inconsistencies" in the voter's overall voting profile on any given issue would be commonplace if the voter votes responsibly.[122] The existence of internal cycles or inconsistencies should not be regarded as a shortcoming of FEFI, for the voter here is modeled not as a "rational-choice" maker, but as a citizen (i.e., a political decision-maker), whose social and political-moral commitments and views (expressed in the first two layers) and personal preferences (indicated in the third layer) could stand in conflict with one another.

The second reason for FEFI to ignore Arrow's Impossibility Theorem is that the theorem does not apply to FEFI, for condition (I) disallows it. In one respect, FEFI is similar to the summation of cardinal utilities, and thus would be brushed aside as unacceptable by the condition (I).[123] FEFI, thus, ignores Arrow's Impossibility Theorem by refusing to acknowledge conditions (CR) and (I) as reasonable criteria of rationality and fairness in social decision-making. As was argued in detail, these conditions are conceptually problematic or narrow, and they disallow too much to be suited for real-life voting situations. Finally, with these two conditions out of the way, one should add that FEFI does not encounter the "problem of indeterminacy" raised by Arrow's Theorem.[124]

It is relevant to recall at this point an earlier argument that, although FEFI is primarily intended to satisfy the *moral* principle inherent in the idea of the citizens' *direct participation* in collective decision-making, it nonetheless can also satisfy some *reasonable* criteria of fairness and accuracy in revealing what might be called the "dominant will" among the citizenry. (As can be recalled from the second part of the preceding section, a more advanced version of FEFI could be characterized as an "*amalgamator-composer*" of "collective wills" based on some reasonable criteria of fairness and accuracy, rather than as a mere "revealer" of the "dominant wills.") Granting to Arrow's Impossibility Theorem all of its assumptions and theoretical tools, the theorem does not prove the impossibility of devising a *reasonably fair and accurate method of composing a collective will* out of the individual wills of voters, which FEFI is designed to do. In light of these considerations, one should conclude, Arrow's Impossibility Theorem should not be seen as a barrier facing FEFI, but rather as an occasion for it to bring to light the now-neglected idea that democracy is primarily a *moral* claim; and make the argument that what Arrow's Theorem *truly* proves is not the impossibility of democracy, but rather the inconsistency inherent in treating democracy as a mere *method.*

Concluding Remarks

This chapter followed in the tracks of the approach laid out in Chapter 1. Toward the actualization of the idea of "democracy as the political empowerment of the citizen" developed there, the chapter focused on the practical aspects of the problem of amalgamating-composing collective wills (or reaching social decisions) out of the wills of the individual citizens. In order to address this problem, the chapter introduced a complex voting scheme called FEFI, and defended it against the theory of voting that raises doubts about the possibility of revealing the public's wills accurately and fairly.

In direct contrast to the theory of voting that takes voters as "rational-choice" makers (i.e., as instrumentally rational seekers of self-regarding interests and aims), FEFI took voters primarily as *citizens*, i.e., as *political decision-makers* whose choice-making is informed and conditioned both by their knowledge of the interests of the public *and* by their political-moral commitments to furthering these interests side by side with their private ones. In supplementing Chapter 1—(which argued that the "conception of democracy as the political empowerment of the citizen" was a "true" and morally justifiable conception of democracy)—this chapter contended that FEFI (a "practical expression" of this conception) can produce collectively decided policies and decisions which could satisfy some reasonable criteria of fairness and accuracy, and at the same time, could lay claim to being meaningful.[125]

This chapter also defended FEFI against those views that question the possibility of democracy itself. William Riker's *Liberalism Against Populism* was taken as a strong representative of these views. The "philosophical disappointment" on the possibility of democracy that Riker bemoans is a product of a systematic attempt to take the foundational political-moral element out of the idea of democracy, and substitute for it the instrumental rationality and technical expertise of a discipline that is quantitative-economic in nature.[126] The irony that permeates the social choice theory's foray into the question of democracy is that, although it starts with positing the individual's choice as a norm that ought to be given an important place (if not the central one) in the theory of democracy, it nonetheless ends up with the sad conclusion that these choices do not really matter, because they cannot be amalgamated fairly or truly.[127] Given these alleged difficulties with the whole notion of amalgamation, as the argument of the social choice theory continues, one then should be content with using the mechanism of voting in a limited capacity, and then only negatively; i.e., in a representative democracy, and only, as Riker puts it, "to control officials, *and no more.*"[128]

In direct contrast to the quantitative-economic approach the social choice theory takes to the question of democracy, the "conception of democracy as the political empowerment of the citizen" regards the question of democracy as *primarily* a *moral* one. It affirms that *democracy is, first and foremost, about the political empowerment and the direct participation of individual citizens in the political process.* It regards the question of democracy as being essentially the question of the people being sovereign and exercising this sovereignty *directly*, i.e., the people (the citizens) being the *actual* decision-makers. As to the *secondary* question of assuring that the citizens' participation is both meaningful and warranted—(in that the "reasonably fair and accurate" scheme of amalgamating-composing their contributions to the collective decision-making process could be found)—the "conception of democracy as the political empowerment of the citizen" puts its faith in technique (both social, and quantitative, i.e., mathematical-scientific). Via arguing for the feasibility of such schemes, the conception shows that the true potential of the idea of democracy has yet to be realized. With this positive orientation toward the collective decision-making enterprise, one can still relish the philosophical optimism about the possibilities of reaching higher degrees of democracy. In light of this optimism, the appeal of the philosophical disappointment about democracy that Riker seems to celebrate fades away considerably.

Finally, one should note that, in addressing the question of arriving at collective or social decisions, the "conception of democracy as the political empowerment of the citizen" opted for a proceduralist approach which stood in direct contrast to the outcome-based approach adopted by the theory of voting. The basic argument of the theory of voting is as follows. Since we cannot assure "good" outcomes (i.e., rationally consistent outcomes as specified by Arrow's conditions), no matter which procedure we employ, we need to lower our expectations of what social or collective decision-making procedures can achieve and only make a minimal use of them. It then follows that we should abandon the search for a "right" set of institutional-procedural arrangements, and instead put our faith in an artificial-extraneous system of social decision-making, hence the representative democracy. Guided by the conviction that democracy is primarily a substantive notion, and that as a moral idea, it requires the direct participation of individual citizens in decision-making, this chapter privileged procedure over outcome. It argued that what matters primarily is that we establish a "right" (i.e., direct-participatory) set of institutional-procedural arrangements. The next step is to assure that the decision-making procedures-techniques we utilize give us outcomes that satisfy some reasonable criteria of fairness and accuracy that could be agreed to by a free, equal, politically knowledgeable, and civic-minded body politic. The last two sections in this chapter argued that such criteria are within the realm of possibility. (As will be argued in Chapter 4, it is possible to rework these criteria in ways wherein they will also be able to satisfy an epistemic requirement.)

Notes

1. As was mentioned in Chapter 1, the term "wills" in this book is used in a technical sense. See note 57 in Chapter 1 for a brief description of the term.
2. FEFI is the abbreviation for "Full Expression and Full Integration."
3. For example, policy option III could boost business at her company and result in a pay raise for her. Option II could help improve the facilities at her daughter's school, and option V might lead to creation of new high tech jobs—her husband is an unemployed software engineer looking for a job.
4. For example, she could be concerned about the deteriorating condition of the infrastructure and the rising unemployment and would see these as the most urgent issues to be addressed by the new budget. This would motivate her to put most of her "particular-public-interest" points on options VI and I, respectively, which she had ignored in the first category, but she thinks would address these particular public concerns better than budget options II-V.
5. For example, she could be of the view that education, healthcare, and elderly-care issues are the most important long-term public concerns. In spreading her points in the "general-public-interest" category, she would consider the long-term effects of each of the six policy options on these concerns and place her 100 points on the one(s) that would help remedy these problems.
6. The voter has exactly $(96{,}560{,}646) \times (96{,}560{,}646) \times (96{,}560{,}646) \approx 9.0 \times 10^{23}$ ways of expressing her will on each voting occasion. To provide a much larger range, voters could be given 1000 points in each category.
7. It should be noted that this version of FEFI is similar to (but *not* the same as) what has been known as the "summation of cardinal utilities method" (sometimes referred to as the "Bentham Method"). The similarity has to do with the fact that both methods rely on performing a summation operation. The difference, on the other hand, lies in that FEFI does *not* find the sum of utilities; but rather, it attempts to evaluate which policy option(s) draw(s) the strongest expression(s) of wills or support from voters. Furthermore, the points assigned to policy options in FEFI do much more than just reflect voters' measurements of personal utilities of those policy options. (Given the complexity of the system, especially with millions of voters participating, such measurements are impossible.) Instead, the points in question allow voters to *express* the intensity of their wills using a numerical system. Moreover, FEFI can also be regarded as a variant of "approval voting" with the added advantage that it accounts for the intensity or the degree of approvals. One can also devise a different version of FEFI that is much closer to the approval voting scheme. For example, each voter could be asked to approve (be it fully or partially), or disapprove, each of the policy options in the same three categories of private-interest, general-public-interest, and particular-public-interest. In this scheme, voters would express their approvals (full or partial) and disapprovals by using a numerical evaluative scale that would allow them to assign a numerical value (0-100) to each of the policy options. It is reasonable to assume that in such a scheme voters would assign in each category a value of 100 points to the policy option(s) they fully approve, 0 to the one(s) they do not approve, and a number in the range of 1-99 to the one(s) they partially ap-

prove, depending on the degree or intensity of each partial approval. The winning policy option again would be the one that gets 50%+1 approval points; otherwise, the highest two approval receivers would face each other in a runoff vote. (It should be noted that this approach would remedy the main disadvantage of the approval voting, viz., preventing voters from expressing the intensities or strengths of their approvals.) Finally, this version of FEFI is similar to the "cumulative voting" scheme that allows each voter as many votes as there are options. For example, each voter is given six votes if there are six candidates or options. In the cumulative voting scheme, the voter has the choice of placing her six votes on just one option, or distributing them among any or all of the six options in any way she deems appropriate. (Lani Guinier of Harvard Law School is a proponent of cumulative voting for she believes this method would help racial minorities to be represented fairly in legislative houses (Amy 2000, p.115).) The original version of FEFI can be regarded as a variant of cumulative voting in which every voter is given 50 points (instead of one vote) for each of the six options.

8. A voter who chooses the protest option could be offered the choice of filling out a questionnaire in order to express specific reason(s) for protesting the ballot. (She would place check marks next to the statement(s) on the questionnaire that best represent(s) her reason(s) for protesting.)

9. One should recall that for Rousseau, the "general will" represents the general/common interest, whereas the "will of all" is the "sum of particular [private] wills." See also note 71 in Chapter 1 for a brief critique of the "general will."

10. If a voter places all of her points in all of the categories on a single policy option, there is a high probability that she is misusing FEFI (i.e., in a mischievous way). Such a misuse reduces FEFI to a two-stage Hare method.

11. Levine puts this "paradoxical situation" as follows: "Popular sovereignty requires that the minority be compelled, and yet popular sovereignty is incompatible with any compulsion" (Levine 1976, p.62).

12. It should be acknowledged that this argument does *not* amount to claiming that FEFI solves the problem of state authority vs. individual autonomy. The problem is irresolvable, except in an imaginary state of what Robert Paul Wolff calls "unanimous direct democracy" (Wolff 1970, p.23). This is how Wolff states the problem: "[t]he defining mark of the state is authority, the right to rule. The primary obligation of man is autonomy, the refusal to be ruled. It would seem, then, that there can be no resolution of the conflict between the autonomy of the individual and the putative authority of the state" (ibid., p.18). For a short discussion of Wolff's argument and the problem it poses for individual autonomy see Weale (1999), pp.65-66. See also Archer (1995), pp.33-34.

13. This is the Rousseauean notion of the individual being both a *citizen* (the one who participates in sovereign authority) and a *subject* (the one who obeys the law) at the same time—discussed in Chapter 3 of *The Betrayal of an Ideal.* "As to the associates," in Rousseau's own words, "they collectively take the name *people*; individually they are called *citizens*, insofar as they are participants in the sovereign authority, and *subjects*, insofar as they are subjected to the laws of the state" (Rousseau 1762, p.22, original italics).

14. Two notes are in order here. First, this argument is similar in some ways to the one advanced by John Rawls in justifying why a "reasonable citizen" should regard the opinion of the majority as a legitimate law, and thus obey it, in a well-ordered constitutional (representative) democracy (Rawls 1999, p.137). Second, the reader should bear in mind that this argument is being presented against the backdrop of the contentions of Chapter 1, including its conceptions of the individual and democratic society, and the assumptions

that citizens are politically educated and believe that the "idealized" scheme is "adequately democratic"—that is, the scheme is a "reasonably fair and accurate" method for making collective decisions.

15. Clearly, this does not add up to an argument for the moral obligation of the individual citizen to obey the authority of the formed collective will, but only that the citizen will have good reasons for obeying it.

16. It is worth recalling in passing the well-known fact that Rousseau regarded this model of sovereignty as a form of slavery. "Sovereignty cannot be represented for the same reason that it cannot be alienated. It consists essentially in the general will, and the will does not allow of being represented. It is either itself or something else; there is nothing in between. The deputies of the people, therefore, neither are nor can be its representatives; they are merely its agents. They cannot conclude anything definitely. Any law that the populace has not ratified in person is null; it is not a law at all. The English people believe itself to be free. It is greatly mistaken; it is only free during the election of the members of parliament. Once they are elected, the populace is enslaved; it is nothing. The use English people make of that freedom in the brief moments of its liberty certainly warrants their losing it" (Rousseau 1987[1762], Bk. III, Ch.15, p.74).

17. Differences between "preferences" and "wills" were discussed in some detail in Chapter 1.

18. One should recall that this question was discussed in Chapter 1 within the context of comparing the notions of "individuated political sovereignty" and "consumer sovereignty." (See also Sunstein (2001a), pp.114-15, and pp.122-23.)

19. A good case in point is the recent claim of "MoveOn.Org," an Internet-based organization of the left-leaning Democrats in the United States. In an e-mail memorandum sent to its members on February 16, 2004, the organization made the following claim: "MoveOn is now over two million people strong in the United States. That's a huge number: the organization we've built together is bigger than the Christian Coalition at its peak."

20. The phrase inside the quotation marks is borrowed from William Riker (1982), p. 167.

21. A good example here would be the case of those Democrats who voted strategically for Senator John McCain in the open primaries in the 2000 presidential elections with the intention of eliminating Governor George W. Bush in the primaries. Most of the strategic voters in this case were of the view that McCain was a greater evil than Bush; nonetheless, they voted for him because they believed that their preferred presidential nominee, Al Gore, would have a better chance of beating McCain than beating Bush in the ensuing presidential elections in November of that year. As an example for the case of "realistic-practical voting," one should mention the U.S. presidential elections in the year 2000. A great number of Ralph Nader's supporters voted for Al Gore because they were certain that Ralph Nader did not have a realistic chance of winning the election and that voting for him would have weakened the chances of the lesser evil, Al Gore, and would have strengthened the greater evil, George W. Bush. The close election results in the state of Florida confirmed what was feared by many Nader supporters. Al Gore could have won Florida, and thus the presidential election, had fewer than 2% of those who supported Nader, voted for Gore. (According to some media reports, over 93,000 Florida residents voted for Ralph Nader. According to a poll, three fourths of these voters would have voted for Al Gore had Ralph Nader's name not been on the ballot. Bush won the election with a margin of less than 1,800).

22. For example, not every gay or lesbian supports the idea of gay-lesbian marriage, for some consider the institution as inherently oppressive.

23. The phrase "public political culture," and the general idea behind it, is borrowed from Rawls (Rawls 1993, p.175). More specifically, the phrase is intended here to denote the positive manners in which the "actual political culture" of society is portrayed in public by the political elites (i.e., the leading political officials and figures, the elite political pundits, the mainstream intellectuals, the leading elements of the mass media), *and* by "self-respecting" citizens. "Public political culture" always presents a positive or ideal portrait of the actual political culture of society and its values. (In Rawls' own notion, the public political cultures of the Western liberal democracies are open to lines of reasoning that take a universalistic approach to political questions, e.g., equal respect for each person and equal regard for the interests of all.) Public political culture does not often coincide with the "actual political culture"—and when it does coincide, it does not do so exactly. For instance, in an authoritarian regime, the actual ground rules of politics might include bribery or rigging of the votes, but such modes of conduct would not be displayed publicly or discussed in the media. The other side of the coin in the "actual political culture" in such an authoritarian regime would be marked by suspicion, pessimism, and distrust of the regime by the populace. In American liberal democracy, members of the Congress do give in to the requests of favors by the lobbyists or corporations in exchange for receiving campaign contributions. Although this is public knowledge, the culprits always deny it; and the leading political elite and media members publicly announce that, despite these unfortunate "rare" occurrences, the system is sound and works well. A third example here would be the case of those private American citizens who harbor racist views against certain ethnic or racial groups; but given the preponderance of the prevailing public political-cultural views on diversity, they do not express these views in public, e.g., when they call radio talk shows, albeit they might discuss these views with close friends or family members.

24. As to the relations between the constitutionally affirmed values and sentiments, on the one hand, and the "public political culture," on the other hand, it goes without saying that an important function of the political elite in each society is to affirm, and disseminate, these values and views throughout society. And if they are successful in this, and if the political system is "healthy" and responsive to pressure from the population, there should be no discrepancies between the elite political culture and mass political culture, on the one hand, and between the actual political culture and the public political culture, on the other hand. See note 23 above for a brief statement on the "public political culture."

25. Burnheim's *Is Democracy Possible?* (1985) places a great emphasis on the importance of lottery in democracy. (McLean (1989) has commented extensively on Burnheim's "statistical representation" concept.)

26. This is how William Riker puts the problem: "the possibility that social choice by voting produces inconsistent results raises deep questions about democracy" (Riker 1982, p.18). On a related note, it should be mentioned that on the authority of Emily Hauptmann's comments that *there are no essential differences among social choice, rational choice, and public choice theories*, throughout this chapter, these theories will be treated as constituting a single enterprise. According to Emily Haumptmann, rational, social, and public choice theories are "closely related" and "sometimes indistinguishable." The difference lies "in emphasis, if not in method" (Hauptmann 1996, p.2). One should also note that some authors, e.g., Knight and Johnson (1994), and Iain McLean (1991), on the other hand, define the social choice theory as a branch of the rational

choice theory that is concerned with aggregating individual preferences into social choices, e.g., Knight and Johnson (1994), p.279.

27. The phrase "theory of voting" is borrowed from Iain McLean who is a public choice theorist. He offers a short overview of the history of this theory in McLean (1991b), pp.176-81.

28. This is how Coleman and Ferejohn characterize social choice theory's problem with voting outcomes (Coleman and Ferejohn 1986, p.11).

29. An outcome is a cycle when in a set of options, e.g., A to Z, A is preferred to B, which is preferred to C, which is preferred to D, . . . which is preferred to Z, which is preferred to A. This problem is named after French mathematician Condorcet who was the first person to study the problem of cycles (McLean 1987, p.26). Also see McLean (1991b), p.178, p.190, and p.195.

30. Riker (1982), p.116, italics added. A more precise and technical statement of Arrow's Theorem will be offered later on.

31. This is a corollary of Arrow's Theorem (Gibbard-Satterthwaite corollary). Iain McLean states this corollary in the following way: "every choice procedure which gives a unique outcome from any given set of individual orderings is either dictatorial or manipulative" (Iain McLean 1991, p.180).

32. Russell Hardin in Craig (1995), Vol.8, p.64.

33. Riker (1982), p.62.

34. Russell Hardin in Craig (1995), Vol.8, p.64. Downs' assumptions were that the voters were primarily self-interested and the candidates were interested in their own elections. For a summary of Downs' view see Cunningham (2002), pp.103-06.

35. This is the "McKelvey-Schofield corollary" stated by McLean (1991), "Forms of Representation and Systems of Voting" (ibid., p.181). McLean adds that "[a] political dimension here means one such as left/right, rural/urban, church/state or Protestant/Catholic" (ibid.). As is commented in Chapter 9 in of *The Betrayal of an Ideal*, the theory of voting takes its lead from Schumpeter's theory of democracy and emulates its overall theoretical orientation and methodology.

36. Riker (1982), p.22 and p.234.

37. Riker (1982), p.11 and pp.234-35.

38. Ibid., p.156, p.167, and p.236.

39. Ibid., p.181, also pp.169-81 and p.237.

40. Ibid., p.113.

41. Ibid., p.238, also p.237.

42. Ibid., p.115.

43. Ibid. p.237.

44. Ibid., p.236.

45. Ibid., p.233. Riker's epistemological skepticism about the knowablity of public wills pushes him down a slippery slope of ontological skepticism about the very existence of public wills, and through that, about whether there exists such social objects as public interests (ibid., p.137 and p.241).

46. Riker (1982), p.113.

47. Ibid. p.115.

48. Ibid., p.234.

49. Ibid., p.11.

50. Ibid., p.239.

51. Ibid., p.9, original italics.

52. As defined by Risse, the multiplicity problem states that "there are several mutually inconsistent but reasonable voting methods" (Risse 2001, p.707).

53. The reader should consult Mackie (2003b), pp.158-72 for a further discussion—and the "debunking"—of Riker's claims about the empirical probability of strategic voting and agenda manipulation. Mackie argues that Riker here mistakes logical possibility with empirical probability.

54. Riker (1982), p.59 and p.234.

55. Ibid., p.59 and p.234.

56. Ibid., p.59 and p.234.

57. Given that outcomes reflect the informed and reasoned judgements of knowledgeable voters, and also given that FEFI is a fair and optimally accurate method of amalgamation-composition, one can further argue that the outcomes produced by FEFI will be "epistemically correct." This latter argument hinges on the validity of the "Generalized Condorcet Jury Theorem" that extends the original theorem to cases in which the ballot contains more than two options or alternatives, e.g., in the case of plurality voting. List and Goodin have argued that this extension is possible, and have laid claim to proving it formally. Expressed in a non-technical and informal language, the generalized theorem states that:

> The correct option is more likely than any other option to be the plurality winner [with k options in the ballot], *regardless* of how likely each voter is to choose *any other* option. Even if each voter is more than $1/k$ likely to choose each of several outcomes, the correct one is more likely to be the plurality winner than any other, just so long as the voter is more likely to vote for the correct outcome than those other outcomes (List and Goodin 2001, pp.286-87).

The notion of "epistemically correct" presupposes the existence of "an *independent standard* of correct decisions," and two other requirements specified by Joshua Cohen's "epistemic theory of democracy" (ibid., p.277n, original italics). Returning to the earlier contention that FEFI produces "epistemically correct" decisions, the assumption is that informed, well-reasoned, and politically knowledgeable voters are more likely to choose "epistemically correct" outcomes than incorrect ones.

58. In contending against Riker's claim (that the results of the theory of voting show that "populist" democracy is unjustifiable), Risse advances a similar argument in defense of Cohen's "epistemic theory of democracy" (alluded to in note 57 above): "different situations require different procedures" (Risse 2001, p.733). According to Risse, the "multiplicity thesis does not undermine the populist conception of voting" (ibid., p.734). (What Risse denotes by "multiplicity thesis" is very similar to claim 1 above.)

59. Many commentators consider Borda Count and Approval Voting as the "optimal" methods, or among the best ones. (Almost all commentators regard Condorcet pairwise comparison method as "the best.") The problem with the Condorcet method is that it does not always produce a winner and that it is "not computationally feasible" for elections with large numbers of voters and three or more options. See List and Goodin for further elaboration (List and Goodin 2001, p.283n). In the case of the Borda Count method, Sarri expresses the view that mathematical investigations show that the Borda Count is the "optimal" method (Sarri 1995, e.g., p.13). Borda Count "appears to be the unique method to represent the true wishes of the voters" (ibid.). (Sarri goes on to say that Borda Count can be manipulated, but adds that as the Gibbard-Satterthwaite corollary has shown, this

is true of all non-dictatorial methods (ibid.)—see note 31 above for a statement of Gibbard-Satterthwaite corollary.) Brams and Fishbum (1983) put forth a strong argument for Approval Voting. As to using Borda Count, the combination scheme can interpret the points entered by each voter in ways that the voter's points profile can be translated into a Borda ranking, and thus perform a Borda Count for all voters' point profiles in order to find the highest two ranked options. The combination method can also interpret the voters' points profiles to make it suitable for the approval voting scheme (e.g., by counting a policy option as approved by the voter if it receives a certain specified number of points from her). See Riker for a concise description of these two methods (Riker 1982, pp.30-31, pp.82-90, and p.95, p.98).

60. According to McLean, "cycles are unlikely as long as the number of plausible candidates is no higher than four" (McLean 1991b, p.195).

61. Those who select the "protest option" in the first round would automatically be forbidden from redistributing their points in the second or the third round.

62. A reference to Riker's view expressed in Claim 5 noted earlier in this chapter.

63. This example is still relevant to the argument, notwithstanding the fact that the present-day U.S. public is neither sufficiently educated nor participatory in political matters.

64. That is "*all*" except the naturally occurring binary alternative voting situations which are "extremely rare." See the discussion earlier in this section.

65. See Riker (1982), pp. 247-49. He mentions Peron by name. Riker's book predates Fujimori's presidential terms.

66. For example, see Pateman's criticism in Pateman (1986), p.45.

67. Riker (1982), p.247.

68. Ibid., p.9.

69. One should be reminded that theories of social choice, rational choice, and public choice are treated as constituting a single enterprise throughout this chapter. See also note 26 above.

70. Hauptmann (1996) provides a review of some relevant literature, pp.22-26 and pp.95-98. She notes that the analogy between market and politics is *formal* and not *substantive* (ibid., p.74). In this, she follows Karl Polanyi. (An example of a substantive analogy would be Karl Marx's economic analysis of the capitalist economy.) Macpherson, to use one example, has argued that "there is nothing in the political system as precise and measurable as price in the economic system" (Macpherson 1973, p.189).

71. Cunningham (2002), p.122. It, as Cunningham continues, "produces a pernicious political culture, it discourages the nurturing of an alternative culture that places civic commitment to public goods and enthusiastic involvement in public affairs at its center" (ibid.).

72. The same can be said about the public choice theory. According to Dennis Mueller: "The basic behavioral postulate of public choice, as for economics, is that man is an egoistic, rational utility maximizer" (Mueller 1989, p.2). As has been argued by many, this *homo economicus* model of the individual is incongruent with empirical evidence on the political and social behaviors of the citizens. Sunstein's argument cited earlier in Chapter 1, as can be recalled, strongly disputed the supposition that individuals behave in self-regarding manners on matters of the public good.

73. Philip Pettit has argued that "[t]he economic supposition may be relevant in some areas of human exchange, most saliently in areas of market behavior. But it clearly does not apply across the broad range of human interactions. . . . I do not call on you in the name of what is just to your personal advantage. . . . I call on you in the name of your

commitments to certain ideals, your membership of certain groups, your attachment to certain people. I call on you, more generally, under the assumption that like me you understand and endorse the language of loyalty and fair play, kindness and politeness, honesty and straight talking" (Pettit 2001, p.82).

74. Hauptmann (1996), p.81.

75. That the identification of "choice" with "preference" is problematic, and unwarranted, is a commonly advanced argument against the social choice theory (and the theory of voting for that matter). In the case of Arrow's Theorem, as will be seen later in this section, this argument is further extended to also include Arrow's identification of "social welfare judgment" with "social choice" (in connection with decision mechanisms).

76. As Amartya K. Sen argues, the traditional economic theory and the rational choice theory identify "personal choice" with "personal welfare" or personal utility. This identification falls apart as soon as one enters "commitment" as a component of choice into the equation (Sen 1977, p.329).. "If the knowledge of torture of others makes you sick, it is a case of *sympathy*; if it does not make you feel personally worse off [or does not decrease your personal welfare], but you think it is wrong and you are ready to do something to stop it, it is a case of *commitment*" (ibid., p.326, italics added). "Commitment does not presuppose reasoning, but," as Sen argues, "it does not exclude it; in fact, insofar as consequences on others have to be more clearly understood as and assessed in terms of one's values and instincts, the scope for reasoning may well expand" (ibid., p.344).

77. Hauptmann (1996), p.86.

78. Ibid., p.85. Hauptmann uses the example of voting on reforming the welfare program in the United States in the 1990s to illustrate this point. Most voters themselves were not using the benefits of the welfare system.

79. The phrase "scientific-deductive" is borrowed from McLean who is a major proponent of the theory of voting and the rational-choice theory. This is how McLean characterizes the enterprise of rational choice: "[t]he rational-choice approach brings scientific deductive methods to bear on politics" (McLean 1991a, p.496).

80. The transitivity property in mathematics: if $a>b$ and $b>c$, then $a>c$; where a, b, c are any three real numbers.

81. John Burnheim advances a similar criticism on the transitivity property as it applies to individual preferences (Burnheim 1985, p.87). It has also been argued that "Arrow conflates 'choice' and 'preference.' This is the source of his problem" (Reynolds 1979, p.355). While preference relations can be transitive, as the argument goes (roughly), the same cannot be assumed about relations among choices, for choices or choice-making are understood to rest on (morally- and logically-based) reasons and not on desires, as is the case with preferences. Imposing transitivity as the only criterion on choice relations would interfere with adjudicating among competing reasons.

82. McLean (1991a), p.508. It thus turns out that the transitivity property is much more than a methodological question, and that it is indispensable to the theory itself, in that it constitutes the core of what is assumed to be the rational individuals.

83. This argument is advanced by Reynolds and Paris (Reynolds and Paris 1979, p.366).

84. Phrases inside the quotation marks are borrowed from Reynolds and Paris who advance this argument (ibid., p.358, original italics). It should be added that the authors' own argument is much more complex than the one presented above. Moreover, one should recall from note 76 above that, as Sen has argued, commitments do not necessarily exclude reasoning. On the contrary, the assessment of one's commitment requires that the scope of her reasoning be expanded (Sen 1977, p.344).

85. The common argument against using intensity information, or cardinal utilities, has been that these utilities are not "observable" or that they require interpersonal comparisons of utility, an idea which is rejected by welfare economics as lacking sufficient scientific justifications (Sen 1995, p.3 and p.7).

86. Barber (1984), p.203. Here, Barber is faulting "liberal democrats" for committing this error. As he goes on, it becomes obvious that he has the theorists of rational, social, and public choice theories in mind, as he mentions Kenneth Arrow, Mancur Olsen, Anthony Downs, James Buchanan, and William Riker by name.

87. Hauptmann (1996), pp.4-5. Although she only mentions rational choice theorists, what she says is equally true of the social and public choice theorists, for she does not see much difference among them. She regards them as being "closely related and sometimes indistinguishable"; they only differ in "emphasis if not in method" (ibid., p.2). Macpherson's overall appraisal of the theory is similar to Hauptmann's. He characterizes the "market conception of democracy" produced by the theory as "self-contradictory" (Macpherson 1973, p.189). That is to say, the theory proves that "if all men act rationally there cannot be a rational democracy" (ibid.)

88. Tangian offers "intuitive, formal, and computational evidence that in a large society [with a large number of voters] Condorcet's paradox (the intransitivity of social preference obtained by pairwise vote) can hardly occur" (Tangian 2000, p.337). Similarly, Levine argues that "[i]n real-world voting situations, paradoxical effects seldom, if ever, emerge" (Levine 1993, p.85).

89. Burnheim (1985), p.87. Burnheim makes these statements in commenting on Kenneth Arrow's Impossibility Theorem.

90. This is how Riker groups the conditions (Riker 1982, pp.116-19).

91. The thirteen possible orderings in this case are as follows: A>B>C, A>C>B, B>A>C, B>C>A, C>A>B, C>B>A, A=B>C, A=C>B, B=C>A, A>B=C, B>A=C, C>A=B, and A=B=C. (Note: symbols ">" and "=" represent preference and absence of preference respectively. A>B indicates that A is preferred to B. A=B means that the voter does not prefer A to B or B to A.)

92. There are varying accounts of these conditions. The ones listed above follow in the tracks of McLean's presentation in the "Forms of Representation and Systems of Voting" (McLean 1991b, p.180). Riker lists six such conditions, the other two being *monotonically* and *nonimposition*. Monotonically basically means the higher the valuation of any option in the individual orderings, the higher should be its valuation in social ordering. Nonimposition, on the other hand, means that social ordering must not contain imposed orderings. A social ordering x is considered imposed if it ends up being the winner, regardless of whether the x was in the domain of possible orderings (Riker 1982, p.117).

93. McLean (1987), p.166.

94. This statement of Arrow's Theorem follows McLean's presentation (McLean 1987, p.174). Riker's non-technical statement of the theorem was provided earlier in this chapter: "no method of amalgamating individual judgments can simultaneously satisfy some reasonable conditions of fairness on the method and a condition of logicality on the result" (Riker 1982, p.116). See Sen for a list of sources on alternative formulations (and proofs) of Arrow's Theorem (Sen 1995, p.1n1).

95. To illustrate this point, one can think of a voting situation in which three voters rank three alternatives A, B, and C. Assuming that all voters prefer A to B, and B to C, and thus A to C, i.e., all list A as their first preference, B as their second, and C as their third (i.e., A>B>C), then the social ordering will list A as the first preference, B as the second, and C as the third (A>B>C). Thus, all conditions of the theorem will be satisfied

with this profile of preference orderings, and the social ordering will *not* be cyclical or intransitive, thus *no* indeterminacy problem. Now, if in the same voting situation, the first voter's preference ordering is A>B>C, the second's B>C>A, and the third's C>B>A, then all conditions cannot be satisfied simultaneously. This profile of voters' preference orderings produces a cyclical result, thus making the voting outcome indeterminable. *It is the existence of this possibility that Arrow's Theorem proves.*

96. One should note that the actual name of the theorem, as Arrow himself called it, is "General Possibility Theorem" and *not* "Impossibility" Theorem.

97. Riker (1982), p.119.

98. Ibid., pp.119-23.

99. Ibid., p.120. To underscore the difficulty involved here, McLean reports on a project that was designed to "calculate the odds [of cycles] for the case of 5 voters and 5 alternatives" that had to be abandoned, for it would have taken 300 hours of mainframe computer time (McLean 1989, p.122n). The project was designed in 1970 when computers were relatively slow. One would assume that the calculation time would be much less if the project were to be tested today. The author does not have any knowledge of recent attempts to calculate the number of cycles for larger cases.

100. List and Goodin note that voting schemes using Condorcet's exhaustive pairwise comparison procedure (which is required by condition (I) of Arrow's Theorem) "are not computationally feasible" for a large number of alternatives in large elections (List 2001, p.283n). They quote Bartholdi, Tovey, and Trick's "Voting Schemes For Which It Can Be Difficult to Tell Who Won the Election" in *Social Choice and Welfare*, 1989, Vol.6, no.2, pp.157-65 as their source.

101. For instance, Tangian claims to providing "computational evidence" to show that in elections with large number of voters, "Condorcet's paradox . . . can hardly occur" (Tangian 2000, p.337). "Simulations and empirical studies," Mackie claims, "corroborate . . . that cycling is an empirical improbability. . . . I have worked through each of his [Riker's] examples only to find that each is mistaken . . . empirical studies show that cycles are of no practical importance" (Mackie 2000a, pp.324-25).

102. Mackie (2000a), p.325. See also Mackie (2003b) for a comprehensive study of Arrow's Theorem, as well as for a complete "debunking" of "myths" perpetuated by Riker's *Liberalism Against Populism*.

103. Some have also questioned the non-dictatorship condition (D). For example, Leslie Green has argued that this condition is "inconsistent" with the assumption of an economically rational individual who "only counts his own costs and benefits" (Green 1983, p.58). Unless he himself is the dictator, Green argues, the individual should not participate in voting, for his costs would outweigh his benefits (ibid.).

104. This is how McLean phrases the criticism (McLean 1991b, p.189, italics added).

105. Friedland (1973), p.54, italics added. To defend (CR) against this contention, the argument has been that without (CR), it would be difficult, as Riker puts its, "to ensure socially satisfactory outcomes"(Riker 1982, p.133). The problem with this argument is that it relies heavily on the transitivity of the ordering of the individual's preferences and does not appeal to those who reject the property as an unreasonable or unrealistic methodological tool. Another way of challenging (CR) is to question Arrow's assumption that "Preference Implies Choice" (DeLong 1991, p.23), as well as questioning his treatment of choice and preference as "synonymous" (Reynolds 1979, p.355). Moreover, as was mentioned in note 81 above, it has been argued that "Arrow conflates 'choice' and 'preference'. This is the source of his problem" (ibid.). While preference relations can be transitive, as the argument goes (roughly), the same cannot be assumed about relations among choices, for

choices or choice-making are understood to rest on (morally- and logically-based) reasons and not on desires, as is the case with preferences. Imposing transitivity as the only criterion on choice relations would interfere with adjudicating among competing reasons.

106. Buchanan (1954), p.116. Here Buchanan works with the assumptions that rational decision-making involves making value judgements *and* that only persons (as individuals) can make value judgements—or that "every value judgement must be *someone*'s judgement of values" (ibid., p.116n10, italics added). According to Buchanan, if we allow ourselves to speak of "social rationality," then we must also allow ourselves to question the wisdom or unwisdom of such rationality. "But does not the very attempt to examine such rationality in terms of individual values introduce logical inconsistency at the outset? Can the rationality of the social organism be evaluated in accordance with any value ordering other than its own?" (ibid.). Buchanan's overall appraisal of Arrow's conditions is as follows: "He [Arrow] fails to see that his *conditions, properly interpreted, apply only to the derivation of the [social welfare] function and do not apply directly to the [actual] choice processes*" (ibid., p.115, original italics). See Sen (1995) for a sympathetic treatment of Buchanan. As Sen reports, Sugden interprets this position of Buchanan as his claim that Arrow's Theorem is a "mistaken attempt to impose the logic of welfare maximization on the procedures of collective choice" (Sen 1995, p.2).

107. DeLong (1991), p.33. A different version of this criticism is expressed by Sugden. According to Sugden, "when a social welfare comparison is being made between two end states, information about how people rank a third alternative can be relevant" (Sugden 1981, p.144).

108. That the intensity or "strengths of preference" are not irrelevant to the process has been argued by a number of commentators, e.g., Sugden (1981), p.144.

109. In the simplest version of the Borda Count, each voter assigns the value of 0 to her least favorite preference, 1 to the preference before the last, 2 to the one before that, and so on and so forth. For instance, if the set of options contains options A, B, C, D, E, and F, and voter x orders them from the first to last as C, D, A, B, F, and E, she will assign the value of 0 to E, 1 to F, 2 to B, 3 to A, 4 to D, and 5 to C. Collective or social ordering will be determined by adding up the values received by each option. The option with the highest number of points is ranked first. In the summation of cardinal utilities method, on the other hand, voter x assigns a utility value to each of the options. The process can be likened to a grading system. Voter x assigns 100 points to her first choice, let us say 82 to her second choice, 78 to the third, 60 to the fourth, 55 to the fifth, and 45 to the sixth choice. Collective or social ordering in this case will also be determined by adding up the values received by each option. As with the Borda Count, the option with the highest number of points is ranked first. Both methods, especially the summation of cardinal utilities method, allow the participants in decision-making to express the intensity of their preferences by assigning higher numbers to their higher or better choices. While the summation of cardinal utilities method works with the expression of the intensity of preferences in an explicit and direct way, the Borda Count accepts the preference intensity information in an indirect, implicit, and limited manner.

110. Arrow (1963), p.27. The common justification for having condition (I), and thus disallowing the Borda Count, has been that the Borda Count is susceptible to manipulation. "If you think a certain candidate is a dangerous rival to your favourite," As McLean puts it, "[you] rank him last" (McLean 1991b, p.189). In response, one can argue that the individual transitive property of ordering used by Arrow's Theorem is also susceptible to the same sort of manipulation. Thus, a rebuttal to McLean's example would be: if you think a certain candidate is a dangerous rival to your favorite candidate, in pairwise com-

parisons, you prefer all other candidates over that candidate, and your favorite over all others; the transitivity property would then automatically rank your favorite first and the dangerous rival last.

111. For example, Sugden, using an example similar to Arrow's, has argued that the "strengths of preference" accounted for by the Borda Count method are not irrelevant to the process (Sugden 1981, p.144).

112. Saari (1995), p.13. Saari goes on to say that the Borda Count can be manipulated, but adds that, as the Gibbard-Satterthwaite corollary has shown, this is true of all non-dictatorial methods (ibid.)—see note 31 above for a statement of the Gibbard-Satterthwaite corollary. In a later work, Saari recommends that Arrow's condition (I) be modified so that it can take into effect the "intensity information" that the Borda ranking provides (Sarri 2000, pp.190-93). As was stated earlier, the common argument against using the "summation of cardinal utilities method" has been that it works with voters' intensity information, and since these utilities are not "observable," or that they require interpersonal comparisons of utility, there is no scientific basis to support the use of the summation of cardinal utilities method.

113. See Tangian (2000) for further details.

114. Sen (1995), p.9. "As Arrow had himself noted," Sen continues, "if the condition of unrestricted domain is relaxed, we can find decision rules that satisfy all the other conditions (and many other demands) over substantial domains of individual preference profiles" (ibid.). See also MacKay (1980), pp.89-103 and Trachtenberg (1993), pp.252-60.

115. Phrases inside quotation marks are borrowed from MacKay (1980), pp.89-103.

116. Trachtenberg (1993), p.258.

117. Ibid.

118. Sen (1995), p.11.

119. In proving the impossibility theorem, as Sen notes, "Arrow had, in effect, stuck to the welfarist structure of traditional utilitarian economics. . . . The conditions imposed by Arrow on social welfare functions had been implicitly used in that tradition" (Sen 1986, p.8). The welfarism of this tradition has to do with the fact that it judges the merits of states of affairs in terms of the "individual utilities" they yield. Moreover, this tradition is consequentialist as it assesses actions, rules, and institutions in terms of the desirability of the results they produce (ibid.).

120. The problem with the standards of rationality in question is that it is identified, as Sen argues, "with . . . justifying each act in terms of self-interests" (Sen 1977, p.342). "There are," as Sen continues, "three distinct elements in this approach. First, it is a consequentialist view: judging acts by consequences only. Second, it is an approach of *act* evaluation rather than *rule* evaluation. And third, the only consequences considered in evaluating acts are those of one's own interests, everything else being at best an intermediate product" (ibid., original italics). It should be added that here Sen is speaking broadly about the standard of rationality that prevails in the economic theory of utility and the rational choice theory, and not about Arrow's Theorem in particular.

121. Friedland and Cimbala (1973), p.61, first italics added, the second original. Having underscored that the proof of Arrow's Impossibility Theorem hinges on accepting a certain value orientation, the authors conclude, "one could easily reach quite a different conclusion from Arrow's proof than that usually drawn. Namely, that it justifies the argument that prevailing utilitarian standards of rationality are inapplicable as guides to action in a democratic society" (ibid.).

122. The reader should review the first section of this chapter, if this statement appears startling.

123. Similarities of FEFI and the summation of cardinal utilities method, as well as their crucial differences, were discussed in passing in note 7 above. As to the rejection of the summation of cardinal utilities method by Arrow's Theorem, Riker argues that the method "gives advantages to persons with finer perception and broader horizons" (Riker 1982, p.118). In Arrow's own view, cardinal representation of utilities stood for what he called "a psychic magnitude" in the mind of the individual, and thus it seemed to him to "make no sense to add the utility of one individual . . . with the utility of another individual" (Arrow 1963, p.11).

124. In fact, this last claim can be established independent of discussing Arrow's Theorem. Given that FEFI tallies the points received by each policy option and sends the top two point receivers to the runoff stage (assuming that no single option receives the qualified majority), the probability of encountering cycles decreases drastically. The only possible type of indeterminacy problem that could be encountered is the case when there would be a tie among the top three point receives. (The same kind of problem can arise when there exists a tie among the top four or top five point receivers, or when all options receive the same exact number of points.) This problem is very improbable when the set of voters is considerably large. One possible way to resolve the problem, in case it arises, would be to use tiebreaker rules.

125. "True" in the sense that the conception remained true to the original meaning of the idea of democracy. "Morally justifiable" in that it both affirmed the value of the positive development of the individual and upheld the primary values of liberty, equality, and community, as was discussed in Chapter 1. Moreover, the collectively made policies and decisions were "meaningful" in the sense that they were preceded by public deliberations among socially and politically educated citizenry, and thus the voting outcomes embodied their reasoned and reflected-upon judgements.

126. The phrase "philosophical disappointment" is Riker's own (Riker 1982, p.200).

127. This is a reiteration of the statement quoted earlier in this chapter from Emily Haumptmann. See footnote 87 above.

128. Riker (1982), p.9.

Chapter 3

A Realistic Democratic Utopia

The conceptualization of the idea of "democracy as the political empowerment of the citizen" presented in Chapter 1 needs to be supplemented with a theoretical-institutional framework within which one can attempt to develop a practical model for the actualization of the idea. The aim of this chapter is to offer such a framework, and at the same time, to develop a practical institutional design for the actualization of the conception of democracy and the implementation of the scheme for collective will-formation that was developed in Chapter 2.

Given the immense size of the modern nation-state, the exercise of the principle of the sovereignty of the individual would be an impossible task without the availability of some potent media of communication, dissemination of information, and the technology of the exertion of the political wills from a distance. It is a contention of this work that such media exist in the present-day American society. As a matter of fact, one reason for insisting in Chapter 1 that the conceptualization of democracy being put forth here was intended exclusively for present-day American society was the existence of an extensive and powerful network of the media of communication and information in this society. These media are nothing other than the high-tech electronic instruments and networks ("*e*-technologies") that in recent years have grown rapidly and penetrated to the very core of almost all aspects of human activity (social, economic, educational, intellectual, and political) in the technologically advanced societies, especially in the United States.

Starting with the fourth quarter of the twentieth century, these media, as part and parcel of the wave of high-tech revolution, have entered into numerous institutions of the American society, and as a result, have altered their organizational structures, as well as affecting their modes of being and functioning. By all indications, it seems that the *e*-technological revolution has just begun. Institution after institution is being won over by this revolution, almost on a daily basis. Banking systems and financial institutions have been completely networked and structured by *e*-technologies. The education system, commerce, economic production and distribution systems, as well as the means and methods of communication, including the print media, are in a constant state of being shaped and reshaped by *e*-technologies. The electronic media have also made significant inroads into the spheres of culture, arts, and entertainment, and are being increasingly integrated into these areas at the same dizzying rate that they are being swallowed up in education and in the economy. The same is true of medicine, scientific research, and intellectual production in general.

The Idea of "*e*-Democracy"

Despite the success *e*-technologies have shown in altering most societal spheres in significant ways in recent years, by and large, they have failed to bring about fundamental changes in the sphere of politics. Notwithstanding this failure, the new technologies of communication and information have found some important applications in this sphere. In recent years, most of the candidates running for political office have resorted to the Internet, electronic town-meetings, e-mail, and fax as media for their campaigns. A good example here is the excellent use Howard Dean's 2004 presidential primary campaign made of the Internet in organizing his supporters and raising funds. This helped establish Dean as the front-runner in the beginning stages of the Democratic Party's presidential primary.[1]

In somewhat similar to the ways the political campaigners have used *e*-technologies to their advantages in recent years, elected officials are increasingly relying on the Internet and similar technologies for their political purposes. Nowadays, almost all major elected officials have their own Internet sites and use them to promote their political views and communicate with their constituents for a variety of reasons, including polling their opinions and inviting comments from them.[2] By all indications, this is a rapidly-growing trend and is generally considered by many as a positive development for democracy in that it expands the access of citizens to the corridors of power. This development has inspired numerous authors and activists to propose innovative ideas for incorporating *e*-technologies, especially the Internet, into politics in this capacity.[3]

Closely related to the sort of usage the elected officials make of *e*-technologies is the idea of "*e*-government." Most of the technologically-advanced countries now offer *e*-government services to their citizens at local and national levels. The most basic, and routinely offered, *e*-government services range from making governmental official paperwork and forms available online and allowing citizens to file their tax forms electronically, to providing online public resources, mainly in the form of information, to citizens. In some cases, these services also make it easier for citizens to "interact" with governments, as they provide opportunities for citizens to express their views on governmental policies and regulations.[4] By many indications, the Canadian federal government seems to be on the cutting edge of expanding *e*-government services.[5]

These three modes of using *e*-technologies in politics are clearly government-centered. Here, *e*-technologies are used as political tools both by the candidates running for government offices, and by those who are already in governments. They are also used by various government offices to render *e*-government services to citizens. Closely related to these modes, there has now developed a fourth category of political applications for *e*-technologies that aims at using them as tools for exerting influence on governments from the outside. Here the underlying idea is to keep governments in check by holding them accountable. In this category, one finds a wide spectrum of ideas, practices, and discourses that are devoted to the premise that the best political use one can make of *e*-technologies is to utilize them for democratic purposes; that is to say, to use these technologies for the purpose of improving the functioning of the existing systems of representative democracy. This conception is often dubbed as "*e*-democracy." Working within this framework, many computer-savvy citizens, activists, and political groups are using *e*-technologies to gain easier access to the elected officials, government offices, and the established political organizations that carry weight in government. For those who possess the know-how, these technologies offer more opportunities than before in the way of influencing the political process, in general, and the decision-making patterns of the government officials, in particular. This conception of the democratic utility of *e*-technologies now seems to dominate *e*-democratic thinking. It is now the conventional wisdom that *e*-democracy can, and does, play a positive role in strengthening democracy. Despite this generally optimistic attitude, however, there are those who are not as sanguine about *e*-technologies' foray into the world of politics. There now exists a vast and rich body of literature on "*e*-democracy" that grows larger by the day.[6]

On the outside of the government-centered approaches to using *e*-technologies in politics, one finds a somewhat different understanding and practice of *e*-democracy. Here one encounters an ever-growing number of Internet sites that devote themselves to promoting alternative types of politics or political ideologies. Some of the sites belonging to this category disseminate the kind of information that they believe is suppressed by governments, or discarded and distorted by the mainstream media. The political goal of these sites is to curb the anti-democratic tendencies of governments and the ruling interests from going overboard by making it difficult for them to control or hide information. The strategy of these sites is to encourage and enable private citizens to widen their sources of information.[7] Some other sites in this category belong to advocacy groups and political organizations that attempt to recruit and organize citizens around their political platforms. These sites work to influence or alter mainstream politics by marshalling support for their causes. They organize online petition-signing, letter-writing, and fund-raising campaigns. In addition, they bring their supporters together to form discussion and strategy groups, and to stage rallies.[8] Some in this category are more interested in combating the politics of the establishment than working within its confines in order to influence it.[9] Still a fourth group of sites in this category promote and pursue special-interest or single-issue politics.[10]

The sixth trend in using *e*-technologies in the service of politics has to do with the mushrooming of numerous Internet sites that intend to influence the voting and decision-making patterns of elected officials by the pressure of the public opinions expressed in them. These sites are seemingly operated by ordinary citizens who invite the public to express their views and cast their non-binding and informal votes on their sites. The argument here is that the elected officials would take notice of the views expressed by private citizens in these sites before making their decisions or casting their votes in the legislative houses. These officials, if they want to be re-elected, as the argument goes, would have to heed the public sentiments expressed in these sites.[11]

Finally, the seventh trend in using *e*-technologies in politics is the idea of "*e*-voting." This idea has manifested itself in recent years in two parallel tracks: one, the idea of casting electronic ballots on the Internet; second, the idea of using electronic voting machines as a replacement for paper ballets in voting booths. What makes the idea of *e*-voting appear appealing to its supporters is its presumed ease, accuracy, efficiency, and speed in the casting and counting of votes. Granting its advantages, however, the opponents of the idea express concerns about the unresolved security issues such as the vulnerability of *e*-voting to fraud and manipulation. In its first manifestation, the idea created great enthusiasm as the Democratic Party of Arizona held the world's first legally binding voting on the Internet during its 2000 presidential primaries. The enthusiasm soon died down as the computer security experts began to publicize the security flaws of the Internet as a voting mechanism. In its second manifestation, however, the idea of "*e*-voting" began to attract considerable attention in the aftermath of the vote-counting snafus in Florida in the 2000 U.S. presidential elections. Numerous states invested heavily in purchasing electronic voting machines. And despite the controversy that surrounded some of them, twenty-nine states managed to utilize these machines somewhat successfully in the November 2004 elections. According to some estimates, up to one third of the voters (approximately forty million) used electronic voting machines to cast their votes on Election Day.[12]

Despite these developments, however, there is ample evidence to support the argument that these modes of integrating *e*-technologies into politics have not altered the overall functioning of this realm in fundamental ways; neither have they necessarily contributed to expanding democracy. The first three modes of using *e*-technologies in politics outlined above, one can argue, have actually expanded the power and the reach of the political and governmental elites over the public. Surely, this by no means can be regarded as a positive development for democracy. In the new *e*-technologies, political elites and governments have acquired new ways and tools for manipulating the public and expanding their political and propaganda powers over the people. Through these media, the elected government officials covertly campaign for their re-election in the guise of communicating with their constituents and soliciting citizens' perspectives on issues. On the opposite end, candidates running for office claim that they use these technologies, especially the Internet, as a means of taking their political message to the people and getting them involved in political discourse, whereas in reality, they are mostly interested in exploiting another set of media for attracting votes and financial contributions from citizens. By many indications, the use of the Internet in political campaigning in recent years has introduced a new sinister element to political image-making and political-marketeering that dominate the fanfare of elections in the late liberal democracy.[13]

As to the idea of *e*-government, it is certainly true that its implementation has made it easier and faster to offer some category of governmental services to citizens. One does not need to drive to a government office to pick up a form, or wait for a day or two to receive it in the mail. All one needs to do is to go to the government site and print a copy by pushing a button and have it there in a matter of seconds. However, the other side of the coin here is that these services have led to distancing government from the people. The everyday citizens are seeing less and less of the government in the form of actual human beings who live and work among them in the community, and more of the government as a distant and alien force existing somewhere out there. This development cannot be regarded as genuinely democratic either.

Insofar as the positive role of *e*-technologies in expanding the democratic control of citizens over governments, or keeping their powers in check, is concerned, it is now becoming increasingly evident that the fourth and fifth modes of implementing these technologies outlined above have empowered only some segments of the citizenry, and in that, only partially and indirectly. As was mentioned earlier, "*e*-democracy" benefits primarily the computer savvy individuals and groups. There are compelling reasons to believe that the traditional political groups, especially those with better or richer resources, and the well-educated, younger, and middle-class individuals are the main segments of the citizenry who are being empowered democratically by the *e*-technologies; that "*e*-democracy" works in "undemocratic" ways. The existing "democratic divide" here results from deeper socio-economic divides that can best be expressed in terms of computer "skill divide," "access divide," and "economic opportunity divide."[14] Among other problems that result from implementing *e*-technologies in these modes, one should call attention to the growing "fragmentation" and "balkanization" of the public sphere that pose serious problems to democracy. This development has been driven, in part, by the citizens' increasing reliance on the Internet as a source of information.[15]

Finally, in the case of *e*-voting, it suffices to state at this point that the mode in which the idea is currently being utilized—that is, primarily as a more accurate and faster way of registering ballots and counting votes—only diverts attention from what is truly democratic about the idea. The democratic potentials of *e*-voting will come to light in the ensuing section in this chapter.

In light of these arguments, it is a contention of this work that the further incorporation of *e*-technologies into the political sphere in the modes and capacities discussed above, no matter how extensive, would fall short of altering politics in major ways. As will become evident in the following section, the prevailing understanding of the idea of *e*-democracy, and the modes in which it is now being actualized, not only limits the true scope of the idea, but also serves to mask the truly democratic potentials of *e*-technologies. So long as citizens do not have a direct input or role in decision-making, their political empowerment would be insubstantial and limited. As can be recalled from Chapter 1, the "conception of democracy as the political empowerment of the citizen" forcefully put forth the idea that substantive democracy at the macro political levels ought to be identified with the direct exercise of sovereignty by citizens. Citizens are truly, i.e., substantially and substantively, empowered, only if they are granted the power to participate in political decision-making in *direct* ways. Given the omnipresence of electronic communication and information technologies in present-day society, and given its overall technological capabilities, it is also a contention of this work that present-day society can use these resources to offer its citizens direct empowerment in politics at macro levels. That is to say, present-day America can enable its citizens to exert their sovereignty directly via implementing *e*-technologies. Simply put, this society has at its disposal all of the technological means needed to enable its citizens to *directly participate* (primarily via a radically different use of *e*-voting) *in decision-making on major legislative and policy matters*—and do this in an efficient and speedy manner.

Without a doubt, the actualization of the idea of the direct exercise of sovereignty via *e*-voting would be contingent upon the existence of some accommodating social-political infrastructures and institutions, and an "accommodating political culture."[16] In order to develop a vision, and a sense, of how this idea could be materialized, and how the accommodating social-political and cultural requirements could be brought about, one would need to delve into a semi-utopian type of theoretical exploration. This will be done in the following section. However, before starting it, one needs to stress at the outset that, far from engaging in pure political fantasizing, such an exploration would be a legitimate political-philosophical undertaking. "Political philosophy," as Rawls puts it, "is realistically utopian when it extends what are ordinarily thought of as the limits of practical political philosophy."[17] "Utopianism," in the words of Jacques Ellul, "does not necessarily mean unrealistic fantasizing. Instead, utopias point toward alternative social models that might be implemented."[18]

With this in mind, what follows is a semi-realistic and semi-utopian theoretical exploration into the form and nature of the institutions that could accommodate the idea of the citizens' direct participation in political decision-making via *e*-voting. This will be done next by means of constructing a theoretical framework that will be referred to as a "Realistic Democratic Utopia."[19] This construction will be followed by a discussion regarding some of the essential features and potentials of this utopia. (It should be stated at the outset that the utopia in question is not intended as a blueprint, but as a rough example and a vision of what is possible.)

Institutional Arrangements of a Realistic Democratic Utopia

The following is an explicit statement about the social-political conditions that would be suitable for the actualization of the "conception of democracy as the political empowerment of the citizen" developed in the preceding chapter. This statement is intended as a description of a system of decision-making that would be a real-life application of the macro principle of sovereignty, i.e., the notion that the citizens ought to exercise their political sovereignty directly. As will be seen, *e*-voting will be the main instrument of exercising this sovereignty. In what follows, there will be no discussions provided regarding the initial social conditions within which these institutions would begin to take shape. Nor will there be any detailed speculations as to how these institutions would be developed or brought into existence. Nonetheless, one can plausibly assume that they would be actualized gradually through successive political reforms and amendments made to the existing liberal-democratic constitution over a period of a few decades. These reforms and amendments would be guided by the commitment to expand the scope of democracy, as well as by some pragmatic considerations that would help guard against some potential social and political problems that could threaten the stability of this society.

One can speculate that the process of the institutionalization of the macro principle of sovereignty would first begin with testing the idea of the citizens' direct participation at local or state levels on minor issues. The process would then be gradually extended to major issues. The next step in the process would be to extend the application of the idea to the decision-making process at national levels, again, starting first with minor issues and then gradually extending it to also include major ones, until the principle is fully realized. As was speculated earlier, the process of actualizing this principle fully could take a few decades. This period is needed in order to iron out the details of the constitutional amendments and their legal ramifications, as well as to develop and spawn the culture of participation and the institutions of public deliberation and political education that would necessarily accompany them. One should reiterate that these institutional arrangements are intended primarily for the actualization of the macro principle of sovereignty. Nonetheless, they could also be mimicked for the development of similar arrangements for those meso and micro levels wherein face-to-face deliberation sessions would prove to be either undesirable or problematic.[20]

Now, one can argue that the institutional arrangements proposed below, far from being utopian, are manifestly realistic in terms of both their technological feasibility and the material resources needed for their implementation. Present-day American society has at its disposal what is needed to implement these arrangements both technologically and economically. Moreover, in light of the state of permanent revolution in the field of development and the utilization of the new *e*-technologies, short of arguing for technological determinism or inevitability, one is hard-pressed to argue that the case for these institutional arrangements is growing stronger as time goes on. Thus, one should argue that the real obstacles in the way of bringing about these institutional arrangements are only cultural and political. These arrangements cannot be actualized unless a sizable majority of the citizenry believes that they could be brought about, and ought to be instituted, and presses the existing political establishment for their actualization. What is needed is a popular political movement devoted to the cause of direct democracy and the political empowerment of the individual citizens.[21]

* * * * *

To begin with, the institutional arrangements, and thus the system of governing that would represent the actualization of the "conception of democracy as the political empowerment of the citizen," viz., the Realistic Democratic Utopia, would be characterized as a *constitutional participatory-representative democracy*.[22] The constitution of this democracy would be a modified version of the existing U.S. Constitution and its major and relevant amendments, including the Bill of Rights. As will be seen, the major difference between these two systems would be that the new system would have two additional elected assemblies: the assembly of the "social-cultural, political, and economic experts" and the assembly of the "trustees of the people." In the rest of this volume, both the "experts" and the "trustees" will be referred to as the "guardians of the public interests," or simply as the "guardians." Unlike the House Representatives and the Senators, the guardians would have no party affiliations. While the "experts" would be nominated and elected mainly on the merit of their demonstrated knowledge of societal issues, and partially on the basis of their integrity (e.g., for being respectable and leading social-political thinkers and philosophers, economists, sociologists, and political scientists), the trustees would be nominated and elected mainly on the basis of their exemplary public-service, their reputations as distinguished and wise citizens who have served their communities (or the society as a whole) superbly, and their reputations as individuals of higher integrity, virtue, and moral standards. The trustees' knowledge of the issues would be their secondary virtue. The main function of the guardians, as well as their *raison d'être*, would be to keep the values of expertise, moral judgement, civic mindedness, and wisdom present in the political decision-making process. The guardians would also be needed to protect the "general interests" of the civil society against the anti-democratic tendencies of the market, on the one hand, and to act as intellectual and moral counterweights against the anti-liberal popular sentiments—which could flare up among citizens in times of crisis—on the other hand.

What follows is a list of the fundamental principles that would underlie the institutional arrangements needed for the realization of the macro principle of sovereignty (the Realistic Democratic Utopia). As was mentioned earlier, the exertion of sovereignty by citizens will rely heavily on using the latest innovations in *e*-technologies including voting electronically, and possibly voting online.[23]

> (I). The existing legislative assemblies of the Senate and the House of Representatives will continue to exist and function as they presently do. The new assemblies (of the "guardians of the public interests"—the "experts" and the "trustees") would work competitively, yet cooperatively, with the members of the House and the Senate, who will be referred to as "representatives" in the rest of this work. The balance of power between these two sets of institutions would be a function of the degree to which citizens participate in voting. The more the citizens partake in voting, i.e., the larger the voter turnout, the more powerful the guardians would be vis-à-vis the representatives.

This fluid balance of power between the guardians and representatives will be considered in further detail below.

(II). There will be a formal distinction between "major" and "minor" issues. Major issues are those that would have larger scope and would impact society in more profound ways than other issues. Examples of major issues would be the national education policy and annual budget planning. It is assumed that there would be about ten to twelve major issues to be decided annually (an average of three quarterly). The remaining issues would be classified as minor issues. Citizens would be encouraged to partake in decision-making on major issues and at the same time would be discouraged from doing so on minor issues. The main responsibility of the guardians would be to engage citizens in discussing and debating the major issues in public before they are put to vote. The political power of the guardians, vis-à-vis that of the representatives, would be a direct function of their success in performing this responsibility. (Higher voter turnout would translate into more power for the guardians.) As to the minor issues, it would be the choice of the individual guardians whether to take part in or abstain from discussing them in public. The main reason for dividing the issues into major and minor categories—as well as for encouraging the populace to abstain from voting on the latter category—is to safeguard the Realistic Democratic Utopia against the dangers that are often associated with the over-politicization of society. As will be seen in Principle V below, any minor issue can be reclassified as a major issue automatically by citizens.

(III). Decisions on major issues will be made jointly by citizens, guardians, and the representatives in a partnership that could potentially assign a much greater weight to citizens' votes than to those of the "guardians" and the "representatives" combined. Moreover, the votes of the guardians can potentially have more weight than those of the representatives. The notion that the votes of citizens can have a greater weight than those of the guardians and representatives combined is contingent upon large voter turnouts by the citizens. In low voter turnout cases, citizens' votes would be weighed less than those of the guardians and representatives combined. This is intended to prevent the small but politically active and vocal special-issue groups or associations from exercising disproportional influence on national politics. The weighing of votes cast by all actors involved (the citizens, the guardians, and the representatives) would be done by carefully crafted, and yet *simple*, mathematical formulas. The guiding principles for these formulas would be to make the weights assigned to the votes cast by citizens directly proportional to the rate of their turnout: the higher their rate of turnout, the heavier the weights of their votes. The following is an example of how the vote-weight assignments could work. Citizens' votes would carry only up to 20% weight if the voter turnout is less than 30%. (The remaining 80% will go to the guardians and representatives, e.g., 40% to the guardians and 40% to the representatives.) If the citizens' turnout would be between 30% and 50%, their votes would carry the weight of 30%. (The remaining 70% would go to the guardians and the representatives; e.g., 35% to the guardians and 35% to the representatives). A turnout of 51% to 60% should assign to the citizens' votes the weight of 40%. (The remaining 60% would go to the guardians and representatives; e.g., 30% to the guardians and 30% to the representatives). A turnout of 61% to 70% should carry the weight of 51% for citizens. (The remaining 49% would go to the guardians and the representatives; e.g., 25% to the guardians and 24% to the representatives). A turnout of 71% to 85% should carry the weight of 60%. (The remaining 40% would go to the guardians; e.g., 22% to the guardians and 18% to the representatives). Finally, for the voter turnout rate of higher than 85%, citizens' votes should count as 70%. (The remaining 30% would go to the guardians; e.g., 20% to the guardians and 10% to the representatives).[24] The question of the balance of power between the representatives and the guardians will be addressed further in Principle XV.

(IV). Decisions on minor issues (i.e., the main bulk of the issues), as well as the legislative decisions regarding the general and day-to-day operations of the government, will be made by the representatives (with some degree of input from the guardians). Moreover, the citizens would not be barred from participating in voting on the minor issues. Instead, they will be encouraged to abstain from doing so in order to prevent the overpoliticization of society. Assuming that only a small percentage of citizens would take part in voting on minor issues, the weight of their votes would be considerably less than those of the representative and guardians (e.g., 3%-10% weight for a voter turnout under 33%). This is necessary in order to assure that small but vocal and politically zealous groups do not exercise disproportionate influence in decision-making.

(V). The distinction between major and minor issues could be blurred if the citizens partake in voting on minor issues in large numbers. If the participation turnout on a minor issue exceeds some specified percentage of points, e.g., 33%, then the citizens' votes should receive a higher weight, e.g., up to 15%. Moreover, if it turns out that a specified percentage of the voters, e.g., 51%, turn out to vote on a minor issue, this would be taken as an indication that the voters regard the issue as a major issue. Consequently, the results of the voting will be declared as null and the issue will be scheduled for re-voting, this time as a major issue. This principle is intended as a measure to keep in check the agenda-setting process and guard against tendencies to manipulate the agenda at the level of the agenda-setting institutions.

(VI). In voting on major issues, citizens would have the option to protest what they might consider to be the "inadequacies" in the list of policy options presented to them, and would potentially have the power to force the agenda-setters to revise the ballot list. If dissatisfied with the ballot, citizens would have the choice to vote for the "protest option" in order to register their dissatisfaction. In the case the protest option turns out to be the winning option—(e.g., if it receives the largest share of points in the primary stage, or receives a certain specified percentage of points, for example, 35%, or makes it to the runoff stage as the second highest point receiver and wins an absolute majority there)—the options will be considered as being "vetoed" by citizens. A "veto vote" would be an indication that citizens did not approve of the policy options presented to them and the agenda-setting commissions must put together another set of policy options based on studying the voting results, and schedule another public vote.[25] This principle is also intended as a measure to keep in check the agenda-setting process and guard against tendencies to manipulate the agenda at the level of the agenda-setting institutions.

(VII). The basic model for collective and direct participation in decision-making (on major issues/policies) would be as follows. Before a vote on a major bill, issue, or policy takes place, for a specified number of weeks or months, there would be ample discussions and informative sessions that would help citizens educate themselves on the issues. These sessions would be directed primarily by the "guardians" (the elected "experts" and "trustees" of the people) both in person (in the local community and townhall meetings) and in *the media* (including in the "electronic townhall meetings," video-conferences, and chat-shows broadcast by various television channels and radio stations that are publicly funded and operated solely for this purpose).[26] During these sessions or in their aftermath, citizens would debate the issues and deliberate both on the virtual plane (whether one-to-one or through "electronic town meetings," a publicly chartered Internet, in "virtual communities," *ad hoc* online discussion groups, and in the media) and in the actual world in public places (e.g., the workplace, associations, local community meetings, "civic homes," churches, synagogues, mosques, schools, colleges, libraries, "local talk shops," and "house parties").[27] Then, at a specified day/time block, citizens would vote electronically (possibly on-line) using their electronic voting cards and PINs.[28] The elected policy "experts" and the "trustees" of the people (the "guardians"), as well as the representatives, would also vote on the issue in their respective assemblies in the same day/time block. The amalgamated votes of citizens would be weighed against the aggregated votes of the "guardians" and the representatives in accordance with a set of carefully developed formulas required by Principle III. According to this principle, higher popular participation in voting would assign higher weight to the votes of citizens vis-à-vis the votes of the guardians and representatives. Moreover, the voting scheme employed would not be the existing monosyllabic yes/no. Rather, it would be the multi-layered scheme of FEFI that would allow each and every individual to "fully express" her wills (both private and public), as well as having these expressions "fully incorporated" into decision-making.

(VIII). The guidelines for the election of the representatives to the House of Representatives and the Senate would remain as they are now. However, the individuals who would be elected to the assemblies of the guardians (the "experts" and "trustees") would need to be fifty years or older, and must be nominated as candidates to these assemblies by citizens (either by popular initiatives or by civic associations), and not by political parties, nor by themselves. Furthermore, unlike the candidates for the House and Senate, the candidates for guardianship would have no party affiliations, nor any connections with market interests. Moreover, both the "experts" and the "trustees" would be elected into their respective assemblies at the two distinct levels of state and the nation—i.e., they would be elected by state and national constituencies, respectively.[29] This would increase the likelihood that "minority" perspectives on "general interests" will also be represented in the assemblies of the guardians. (A modern electronic version of Thomas Hare's voting scheme could be used for electing guardians at national constituency levels.[30])

Finally, the guardians would serve a six-year term and could be re-elected.

(IX). The election campaigns of the guardians would be funded by the public. As to the representatives, their election campaigns would be supported jointly by public funds and their parties.

(X). The status and power of the presidency would be reduced to a ceremonial one. The president would be the head of state and not of the government. The decision-making power in emergency cases, when quick actions and responses to an unforeseen and sudden event were needed, e.g., a sudden invasion by a foreign power, would be left to a "High Council" of guardians who have been elected to their assemblies on national tickets and who have been elected to the High Council by their fellow guardians. The council will consist of seven individuals who would elect a guardian among themselves as the highest-ranking guardian.

(XI). The agenda for public voting can be proposed by any member of the citizenry, civic associations, guardians, and representatives or their political parties. However, given the influence and respect the guardians would command, they would have more success than others in pushing forth the agenda they deem important. Approval of the proposed agenda for public voting would require a complex process of debate and deliberation, as well as an unavoidable degree of political maneuvering and influence peddling.

(XII). Given that this system of governing is designed on the basis of the commitment to maintaining and expanding democracy, the constitution of the Realistic Democratic Utopia would include amendments to facilitate the realization of this end, including an amendment to guarantee the easy access of all citizens to the technological equipment needed to participate in voting electronically, as well as amendments to offer the highest feasible level of economic equality without restricting property rights in a major way.[31] Finally, the constitution of the Realistic Democratic Utopia would also contain a set of protective clauses that would severely frustrate attempts to dismantle direct democracy or weaken the individual rights and liberties constitutionally guaranteed to all.[32]

(XIII). Structures and schemes of decision-making at the state and municipal levels could resemble those used at the level of the national/federal government. At the micro levels, especially at those of the workplace and the local community, decision-making schemes could combine these models with the potentially more direct forms of deliberative participation (e.g., face-to-face) in organizing the discussion sessions and making decisions.

(XIV). The Supreme Court would continue to exist with the same degree of power and significance it now has. The Supreme Court would have the power and authority to strike down the decisions that might be challenged as unconstitutional, if it finds them so. Judges to the Supreme Court would be nominated by the High Council of the guardians. These nominations would then be confirmed by the guardians and representatives, and would finally be put to the citizens for ratification. Supreme Court judges would serve an eight-year term and could be re-nominated for a second term.

(XV). The main distinction between the guardians and representatives can be conceptualized as the contrast between reason, morality, and virtue, represented by the guardians, on the one hand, and the this-worldliness and materialism of life and market relations, represented by the representatives, on the other hand.[33] Moreover, the guardians would not have any party affiliations, nor any connections to market interests. They can have affiliations with civic institutions. The representatives, on the other hand, would have party affiliations and would be assumed to be susceptible to lobbying influences, as they are in present-day society. Furthermore, they would be individuals who would be driven both by their commitment to public service, and by their own self-interests and ambitions for power and status. Moreover, while the guardians would guard the public interests, which broadly speaking, are the long-term and "general interests" of all citizens within broad national scopes, representatives would attend the economic interests, which in most cases, are particular interests or "special interests" and have local-regional scopes, as well as direct or immediate impacts on the lives of the citizens involved. (Placing this contrast between the functions of the "guardians" and "representatives" within the context of the discussion of representation in *The Betrayal of an Ideal*, it becomes clear that while the former would function in ways similar to the "Burkean" type of representatives, the latter group would conduct themselves as representatives in the "Madisonian" sense of the term.[34]). Finally, the balance of power between the guardians and representatives would be a fluid one. One major responsibility of the guardians would be to engage the citizens in taking interest in major issues, educating the citizens on these issues, and engaging them in debates. The weight of their power (i.e., the weight of their votes), vis-à-vis the power of representatives, would be directly proportional to their success in fulfilling this responsibility. The more they succeeded in engaging the citizens in debates, participation in discussions and, ultimately, in voting, the more their own votes would weigh vis-à-vis the votes of the representatives.[35]

* * * * *

These are taken as the necessary and fundamental principles for the Realistic Democratic Utopia that would facilitate the direct exercise of sovereignty by citizens. The following section will first provide a description of the actual process of voting electronically in the Realistic Democratic Utopia, and then will offer a discussion regarding some of the potential objections that could be raised against the Realistic Democratic Utopia, as well as offering a discussion regarding some of the most essential features of this ideal democracy.

Citizens' Participation in the Realistic Democratic Utopia

Before going any further, it should be stated at the outset that the notion of voting electronically does not necessarily imply voting on-line. No doubt, voting on-line would make the practice of voting simpler and thus increase voter turnout and the citizens' direct participation. Notwithstanding this great benefit, there are a number of problems associated with voting on-line that need to be discussed. This will be done in Chapter 4. For the time being, it suffices to mention that by voting electronically, here, is meant a scheme of voting that relies heavily on the latest *e*-technologies, such as fast computers, electronic voting machines, and the electronic means for dissemination of information. Now, one can argue that the multi-layered scheme of voting presented in Chapter 2 (FEFI) would be an appropriate scheme for voting in the Realistic Democratic Utopia. This scheme allows voters to express their wills fully, i.e., in virtually infinitely many ways. Moreover, the scheme fully integrates the expressed wills of the voters into the process of composing the collective will. The following example will be helpful in developing an understanding of the utility of this scheme and the actual workings of the Realistic Democratic Utopia.

Assume that the question of the budget plan for the upcoming year will be put to the public vote as a major issue. Furthermore, assume that a national committee composed of some guardians, representatives, economists, and citizens, has completed a study and has developed a set of six budget options to be considered. These options offer varied and meaningful choices for citizens to consider. (Another example would be the issue of legalizing gay-lesbian marriage that was discussed in Chapter 2.) Eventually, only one of these six options will be chosen.

Before the actual voting takes place, there will be three months of well-organized debates and discussions around the topic. The guardians will lead these discussions in the printed media, and in the community-based and nationally-televised electronic town meetings. The latter meetings will be broadcast by television stations that are publicly funded and operated specifically for this reason. Citizens will partake in electronic townhall meetings by calling in, sending e-mails, or by video-conferencing. The main purpose of these discussions-deliberations will be to facilitate the education of the public on all of the budget options presented to them. Given that the guardians will not have any party affiliations or ties to economic interests, their educational campaigns will be regarded as impartial, despite the possibility that some guardians might choose to declare their preferences for one or more of these plans. In addition to these discussion groups, there will be numerous other sources of information for the public. Representatives and their parties, as well as civic associations, and special interest groups, will provide their own partial views on each of the budget options, and will promote their preferred plans. Citizens, in addition to taking part in the local and nationally televised townhall meetings, and educating themselves using other sources, will also discuss the plans among themselves within the associations or organizations they are affiliated with, and at the local townhall or community-based meetings, as well as at various social settings and informal get-togethers.

During the voting time block in the day of voting (e.g., a time block of twelve hours), all eligible voters will exercise their voting power. They will use their computers at home or at work, or electronic voting machines placed in secured (and staffed) rooms at convenient public locations such as libraries, schools, colleges, and community centers to enter their votes.[36] The guardians and representatives will also vote during this voting block in their respective assemblies. If, at the end of the day, one of the budget options wins a qualified majority (e.g., 55%), then the issue will be declared as decided; otherwise, there will be a runoff vote between the top two vote-receivers during the following week or the next voting session.[37] Voters will have the "protest option" in the FEFI scheme as their seventh option to choose, in case they are dissatisfied with the budget options on the ballot. If a voter believes that the budget options listed on the ballot fall short of offering her a meaningful and varied set of options to choose from, she will place all of her voting points in one, or two, or all of the categories of private-interest, particular-public-interest, and general-public-interest on the seventh option in order to protest the ballot. At the end of the voting time block, the amalgamated protest points will be sent to the agenda setters as feedback to be studied for policy-option-making purposes in the future. The protest option can also function as a "veto vote," if it meets a certain set of requirements.[38] The main purpose of the "protest option" is to keep in check the agenda manipulation tendencies in the agenda-setting institutions, on the one hand, and to keep the agenda setters attuned to the sentiments of the public, on the other hand. The "protest option" is also intended to discourage and minimize non-participation that might result from some citizens' dissatisfaction with ballot choices.

This example provides an insight into the most essential feature of voting electronically in the Realistic Democratic Utopia. The remaining features of the utopia will come to light within the context of the ensuing discussions that will focus on some of the objections that could potentially be raised against the Realistic Democratic Utopia.

* * * * *

To begin with, to the potential charge that "too many policy options to choose from, too many expressed views, and too much talk during three months of discussions and deliberations could confuse the voters on what to vote for or how to vote," the response would be that, to the contrary, this would actually be a good thing in itself. This is mainly because the voters do not need to, and perhaps should not, have a clear position on the issue put to vote. The actual moment of voting would be the true moment of decision-making for voters as they would consider the questions posed to them on the ballot. The voter would answer these questions carefully and express her views as they really are, with her certainties and her doubts, as she would bear in mind that every point of her decision-points would count. In using FEFI, the voter would have 100 points to vote (or to devote) in each category (a total of 300 points for each voting occasion). In each category, the voter would spread her 100 points among the six policy options in any way she sees fit. In answering the question in the private-interest category, she would think in terms of which policy option(s) would serve her private-interests and spread her 100 points accordingly. In answering the particular-public-good question, she would think along the lines of which policy option(s) would serve those interests of the public that are directly related to the issue in question (e.g., the national education policy), and would spread her points accordingly. Finally, in answering the question in the general-public-good category, she would think in terms of how these policy options could benefit the public's long-term interests in the general scheme of things and place her 100 points according to her thinking. Once done with voting, she will walk away from the voting machine thinking that *she has fully expressed her wills on the issue in question, even though she did not have a clear position on it*. Once the voting time block has expired, the "democracy computer" would tally the votes and present the result to the citizens of the Realistic Democratic Utopia as their "collective will" composed of their individual wills.

No doubt, viewed from the perspective of the dominant elitist political culture, the institutional arrangements proposed above, and the system of the citizens' direct participation in decision-making that they are intended to actualize, appear ambitious and somewhat utopian. The prevailing elitist political culture is overwhelmed with an enormous sense of mistrust of the ordinary people's moral, political, and intellectual competence. The history of this mistrust in the tradition of Western civilization goes as far back as ancient Athens—Plato being one of its major proponents.[39] The question of the political and moral competence of ordinary citizens, as well as the question of the extent to which the idea of the ordinary citizens' participation in direct decision-making is utopian, will be discussed in greater detail in Chapter 4. For the time being, it suffices to mention that, ironically, a main feature of Plato's own theory of government will be used in Chapter 4 to dispel the elitist mistrust of ordinary people that he himself championed. Moreover, the same feature will be used in arguing that the idea of direct participatory democracy is more realistic than utopian. As has become apparent by now, the assemblies of the guardians in the Realistic Democratic Utopia are Platonic in some respects—notwithstanding the fact that the educational functions assigned to them are not.

Now, what is realistic about the Realistic Democratic Utopia is that the technological and material conditions for the actualization of its institutional arrangements and legislative structures are already at hand in present-day society. The system of democratic governing presented above depicted only the fully operational stage of the Realistic Democratic Utopia. The project is envisioned to be realized in stages, and could take decades before the fully operational phase is achieved. It might well be the case that the actual process of constructing the utopia will lead to the realization that the vision of the Realistic Democratic Utopia is not as realistic as it was originally assumed, and that it can be materialized only partially. A much more realistic scenario would be to leave out altogether Principle I, the assemblies of the guardians, in the beginning stages and assign the functions of leading discussions and educating the citizenry to the well-known and respected leaders of civic associations. To function in this capacity, these leaders would need ample local and national exposure in publicly funded and publicly operated media.[40] Given that reforming constitutions in major ways is a difficult undertaking, it is very likely that the first principle could be the last one to be realized.[41]

Another reason for referring to this democratic utopia as realistic is that the process of its actualization will be guided by pragmatic considerations, prudence, and trials and errors, in addition to being guided by the moral-political principles laid out in Chapter 1. One can take Principle III as an example. The weight assigned to the votes of the citizens in terms of percentages stipulated in this principle could be lowered in the beginning stages, and then be gradually increased over the course of a few years. The principle of formal division between minor and major issues (Principle II) is another one of these pragmatic considerations. The main idea here is to prevent the over-politicization of society that could prove counter-productive to the cause of furthering democracy. The history of the modern world is replete with evidence that over-politicization undermines stability which in turn puts democracy in jeopardy. Another practical consideration is the efficiency. Too much deliberation, too many issues to deliberate about, and casting too many votes, could take "too many evenings."[42] Many citizens may not want to be involved in politics extensively. However, as Principle IV stipulates, those citizens who want to have greater involvement in social decision-making will not be barred from doing so. Moreover, Principle V leaves the gate wide open to an unlimited participation of the citizenry in decision-making by allowing them to turn any minor issue into a major one, in case they choose to have a greater share in decision-making on it.

In passing, one should point out that the decision-making process on the major issues in the Realistic Democratic Utopia can also be construed as a system of "ratifying" the policy options that are drafted by the agenda-setters (mainly guardians and the representatives) and put before the public for voting. In this sense, the citizens' direct participation in decision-making conforms to Rousseau's famous proclamation and requirement that "[a]ny law that the populace has not ratified in person is null; it is not a law at all."[43]

One should stress that FEFI is proposed primarily for the direct exercise of individuated sovereignty and, thus, for macro levels of the state and the nation; it is mainly intended to remedy the difficulty of assembling all of the citizens in one physical place at the same time. Nevertheless, one can make the case that the scheme can also be utilized at meso and micro levels when face-to-face interactions prove to be impractical, counterproductive, or undesirable. When, for example, the discussion or deliberation sessions would become too difficult to manage, or when the past experiences of gatherings of a certain group of individuals would indicate that face-to-face interactions could create a hostile environment and, thus, decision-making could become an excruciating experience for the participants, one then can turn to a variant of FEFI for decision-making. In these circumstances, before the actual electronic voting would take place, the group members or the would-be participants would be given the opportunity to hold indirect discussions with each other for a number of days using an impartial medium such as an Internet site that has been devoted specifically to the group or to the voting occasion. Each participant would express her view on this site—(even anonymously, if they choose, an option that is not available in face-to-face interactions). Once the specified deliberation-discussion period ends, the decision in question can be made by voting electronically or by having the group members assembled to cast their votes by secret ballots.

One can anticipate that those who regard face-to-face interactions among citizens as the main feature of participatory democracy, and who assign a greater value to it than to exercising sovereignty directly (e.g., Pateman and most theorists of deliberative democracy), would take issue with the electronic voting scheme for lacking the face-to-face component. In response, one can argue that some of the benefits often associated with the idea of face-to-face interactions, save their community-bonding effects, are largely overrated. Leaving aside the fact that face-to-face deliberation sessions could be time-consuming, if not managed well, they can also be divisive and can tear communities apart instead of bringing them together. The situation depicted in the example above is a case in point. There is always potential for abuse in such gatherings.[44] These difficulties aside, moreover, one neither has good reasons, nor convincing historical evidence, to believe that citizens would want to participate in face-to-face interactions for political decision-making purposes. The greatest historical evidence for face-to-face participatory decision-making, i.e., ancient Athens' experiment, lends support to this claim. According to Ian Budge, even though the size of the eligible citizen body in Athens was 20,000-30,000 (not including slaves, women, and children), the "actual attendance [was] much less, at most 6,000—even though citizens were paid to attend."[45] One should also add that at the time, decision-making efficiency and saving time was not a major concern of society; and a good portion of the attendees, thanks to their slaves, did not have to perform any sort of productive labor and thus, had ample time to spend in these sessions.

Furthermore, those who assign greater importance to face-to-face interactions for its educational benefit overlook the potential danger that such sessions could be frustrating to some participants and thus, could stymie the educational process. By contrast, one can argue that deliberative and discussion sessions, such as the ones to be led by the guardians, could be educational, and help citizens in forming educated and informed views on the issues before they cast their votes. Joshua Cohen, one of the most prominent proponents of deliberative democracy, acknowledges the educational benefits to citizens when the political "elites" debate or discuss the issues among themselves.[46] Despite these problems with face-to-face deliberation sessions, one should acknowledge their great educational utility, when they work as intended. One should also stress that the "conception of democracy as the political empowerment of the citizen," with its decision-making scheme (FEFI), is not inimical to the idea of face-to-face dialogue, but only that it is skeptical about the efficiency and productivity of conducting politics in this fashion, especially when these sessions are required to end with decisions on the spot. Notwithstanding this skepticism about face-to-face dialogue, one should also emphasize that the electronic voting scheme being advanced here is premised on the educational utility of the public dialogue. As discussed earlier, the scheme *requires* that a lengthy public dialogue (e.g., a few months) be held *before* the actual voting takes place.

An important feature of the Realistic Democratic Utopia is that therein the role of political parties would be greatly reduced in comparison to the power they now command in present-day American politics. This reduction in their power would not be mandated constitutionally, as is the case with the weakening of the power of the presidency (Principle X), but rather would follow as a natural consequence of the diffusion of the loci of power throughout the society. There exists ample empirical evidence to suggest that the political parties are steadily losing ground insofar as the question of membership loyalty is concerned.[47] Citizens who are registered party members are now thinking and voting more independently than ever.[48] One can assume that the further weakening of the power of existing parties would lead to the formation of new smaller parties.[49] The greater role assigned to the civic associations in the Realistic Democratic Utopia, as well as the allocation of some of the powers to guardians, would multiply the sources that formulate and disseminate coherent and grand conceptions of the common good and social-political doctrines—a function that historically has been a birthright of political parties.[50] The new smaller parties and the strengthened civic associations in the Realistic Democratic Utopia would also function very much the same way the large parties do now insofar as the questions of organizing political activities or rallies and building consensus are concerned. An important consequence of the weakening of the power of the existing large parties would be the diminishing of incentives for voting strategically or ideologically. No doubt, a great source of the power of political parties in present-day society is their financial resources. This aspect of their power will be rivaled and offset by having the guardians' election campaigns and educational functions funded publicly. Moreover, the waning power of the political parties in the Realistic Democratic Utopia would have the effect of weakening the existing patterns of partisan blocs in the legislative assemblies, and would lead to the emergence of makeshift or *ad hoc* blocs and fluid factions in these assemblies. However, given that in decision-making on major issues a large portion of the inputs entered into the process would come directly from citizens, these blocs and factions would have significance mainly in voting on minor issues.

While on the subject of the legislative assemblies, it would be appropriate to discuss the *representative* component of the government in the Realistic Democratic Utopia. Recalling from Part II of *The Betrayal of an Ideal*, the negative evaluation of the representative form of government offered there was not intended as an outright rejection of the idea itself, but as a critique of the *form* it assumed under the liberal and liberal-democratic states. Briefly stated, this form of representative government had two main shortcomings. It was elitist by design and kept ordinary citizens away from the policy and decision-making process. Moreover, it was susceptible to influence-peddling by the "economic society" in general, and to the social strata that controlled or commanded it, in particular. (These two shortcomings were characterized later in Part III of the work as "audience democracy" and "fund democracy," respectively.) Now, the guardian component of the representative government in the Realistic Democratic Utopia is designed in part to specifically address these two shortcomings of the liberal-democratic form. First, one of the main responsibilities of the guardians is to promote the participation of citizens in decision-making (as well as preparing them for this participation by facilitating their political and civic education, and also by generating "social capital")[51]. Second, the other main responsibility of the guardians is to counteract the anti-democratic tendencies that the "economic society" has exhibited historically. As the custodians of the "general interests" of society, the guardians would assure these interests are represented as well as (if not stronger than) the "special interests"—which, given their abundant financial resources, have historically been represented by the representatives.[52]

* * * * *

The rest of this chapter will be devoted to addressing some areas of concern that stem from apprehensions about the direct involvement of ordinary citizens in political decision-making. Recalling from the discussion of the justifications for the representative form of government in Part II of *The Betrayal of an Ideal*, historically, the idea of the ordinary citizens' direct involvement in decision-making has been regarded as imprudent, for fear that it could lead to disastrous consequences if actualized. Given that an in-depth examination of a full range of such concerns will take place in Chapter 4, in what follows only two such concerns will be addressed, and then only to the extent that they directly relate to the public decision-making process in the Realistic Democratic Utopia.

One of these concerns is the fear that the citizens of the Realistic Democratic Utopia could be misled to vote democracy out of existence by using the immense democratic power that FEFI places at their fingertips. "What if," one might ask, "the citizens become mesmerized by a charismatic guardian who would turn into a demagogue and would lead the citizens to vote to dismantle the utopia and elect him or her as their lifetime dictator-leader?" In addressing this fear, one can argue that the depicted scenario does not seem very likely for a variety of reasons. For one thing, as was stipulated in Principle XII, the constitution of the Realistic Democratic Utopia would contain clauses that would frustrate attempts to dismantle direct democracy.[53] Moreover, the system of checks and balances, the distributed and multiple loci of power located in civic associations, parties, and individual guardians, as well as the existence of a diversified field of political views and interests, one can argue, would prevent demagoguery from going overboard. Lastly, one could respond that, by the time the realistic utopia has become fully functional, the citizens will be politically educated and sophisticated enough not to fall for demagogues. This problem will be revisited in Chapter 4. For the time being, it suffices to argue that demagoguery is a much greater danger in the present-day United States than it would be in the Realistic Democratic Utopia. In the circumstance characterized by the low political sophistication of the citizenry and apathy toward politics, a well-financed would-be demagogue could appear from nowhere and make significant inroads into the existing political system. One can cite Ross Perot's presidential campaign in the 1992 U.S. presidential elections as a good case in point here.

The second area of concern would be the problem of the "tyranny of the majority." To this concern, one should respond that, for the reasons just cited, this is not a likely scenario either. That is to say, the system of checks and balances, the distributed and multiple loci of power, and a diversified field of political views and interests would get in the way of forming a tyrannical majority that is intent on finding ways "to invade the rights of other citizens."[54] Moreover, the problem of the tyranny of the minority—(when an activist, well-organized, and ideologically-biased minority of citizens attempts to impose its wills on others)—is out of the question. The only way for a minority to tyrannize the Realistic Democratic Utopia would be to hijack the agenda-setting process or to find ways of influencing it disproportionately. But, this threat can be countered easily by the rest of the citizens through exercising their "veto votes" to throw out the policy options dished out by a tyrannical minority.

Finally, before closing this section, one needs to mention in passing that the idea of using the modern electronic media of communication in the service of democracy is not a new concept at all. This idea has been around for as long as the technologies that support it have existed. The third chapter of Iain McLean's *Democracy and New Technology* is devoted to a survey of numerous democratic-friendly electronic systems and technologies that have been in existence since the late 1970s.[55] And as Ian Budge has noted in his survey of the literature on the feasibility of using modern technologies for expanding democracy, these sorts of technologies "have been in place for a long time."[56] A good example here would be the system proposed by Robert Paul Wolff as early as 1970. His "system of in-the-home voting machines" would use a device "attached to the television set which would electronically record votes and transmit them to a computer in Washington." [57] "In order to avoid fraudulent voting," he suggested, "the device could be rigged to record thumbprints."[58] The main shortcoming of Wolff's electronic voting scheme, as he himself acknowledged, was that the voters were passive in the debate. This problem of passivity has largely been solved in the last decade or so by the rise of interactive technologies such as e-mail, chat-rooms, and video-conferencing.

Concluding Remarks

This chapter attempted to develop a theoretical framework that was intended to serve as a theoretical model for the actualization of the "conception of democracy as the political empowerment of the citizen." The Realistic Democratic Utopia presented above was such a framework. What is truly novel about the Realistic Democratic Utopia is the way in which it uses the idea of *e*-democracy to revive the moral content of the original idea of democracy.[59] In doing so, the Realistic Democratic Utopia pushes the idea of "*e*-democracy" beyond the limits of the prevailing understanding and modes of its utilization which, as argued earlier, only serve to mask the truly democratic potentials of the idea. What is also novel about the Realistic Democratic Utopia is that it preserves the best of the existing representative system, and at the same time, remedies its shortcomings by introducing the notions of the guardianship of the public interests and the citizens' direct participation in decision-making. Here the idea of "*e*-democracy" lives up to its truly democratic potentials.

Recalling from the closing pages of Chapter 1, this work is primarily concerned with the question of direct-participatory democracy at the macro levels (and particularly at the level of the national government). For this reason, the exclusion from a detailed consideration of the question of direct democracy at micro levels in the Realistic Democratic Utopia, especially at the local community and the workplace levels, should not be interpreted as overlooking the importance of direct democracy at these levels. Given the predominance of the face-to-face interactions among citizens at smaller micro levels, deliberative modes of discourse and decision-making would acquire more relevance and applicability in practicing direct democracy at these levels.[60] The Realistic Democratic Utopia cannot be truly democratic if it ignores the question of direct participation at the workplace and the local community levels. These two constitute the core components of any model of direct-participatory democracy, in that they furnish the actual social settings for developing communal bonds and acquiring hands-on civic and political education. As Principle XIII hinted above, the assumption here has been that the practice of democracy at these core levels in the Realistic Democratic Utopia would potentially be more direct and more participatory than the practice at the macro levels. Given that there already exists a rich literature focused on the question of direct-participatory democracy at the workplace and local levels, this work has chosen the macro level of the national government as its primary area of concentration.[61]

Notes

1. John Kerry's 2004 presidential campaign also showed great success in raising funds using the Internet. This success helped Kerry's campaign to quickly narrow the election fund gap between himself and George W. Bush. Another good example is Jesse Ventura's successful campaign for governor of Minnesota in the 1998 elections that made an effective use of fax machines, cell phones, e-mails, and a website.

2. A good example here is the following e-mail message the author received from Congresswoman Stephanie Tubbs Jones on July 22, 2004:

> Dear Majid,
>
> As your Representative in Congress, it is my job to serve you in the most timely and effective manner. To that end, it is important for me to know where you stand on issues. It is also important to inform you about legislation and activities in Washington, D.C. and their potential effect on the Eleventh District of Ohio.
>
> Today I am launching a new communications tool to send occasional Email updates to constituents. . . .
>
> Technology has enabled many of you to be in contact with me. This Email update will open new channels of communication with you. . . .
>
> . . . I look forward to hearing from you.
>
> Sincerely,
>
> Stephanie Tubbs Jones
> Member of Congress

3. Steven L. Clift's online article titled "E-Government and Democracy: Representation and Citizen Engagement in the Information Age" is an excellent point in case here. The article can be accessed on the Internet at <http://publicus.net> (17 Mar. 2004). See also the site for "Minnesota e-democracy" at <http://www.e-democracy.org> (14 Sept. 2004) for similar perspectives.

4. As an example, see Steven L. Clift's online newsletter at <http://do-wire@lists.umn.edu> (17 Mar. 2004) for a list of the latest services offered to citizens by the state government of Nebraska. Clift's online article mentioned in note 3 above is a strong statement on the potentials of *e*-government. Clift argues that some of these potentials have already been utilized, while others remain unexplored. The article offers ideas and strategies for expanding *e*-government. Among these one should mention using *e*-technologies to expand "citizen-to-citizen discussions online," to enable officials to communicate and interact with citizens (to send e-mail notifications to citizens, to "consult" with them and/or receive input from them), and to introduce information to the policy-making process.

5. For further details, visit <http://www.crossingboundaries.ca> (21 Feb. 2004). See also Allison Taylor's article "Canadian Government on Top When Developing e-

Government Services" on site <http://www.gcis.ca/cdne-478-apr-10-2003.html> (21 Feb. 2004). Moreover, see Jeffrey Roy's "E-Government in Canada: Meeting the Digital Imperative" in McIver and Elmagarmid (2002).

6. A quick review of the literature on the use of the latest *e*-technologies (especially the Internet) for democratic purposes ("*e*-democracy") shows that there are two broad trends on the question. The first trend is cautious and by far seems to represent the prevailing school of thought on the question. The authors representing this trend, although they acknowledge the democratic possibilities opened up by *e*-technologies, doubt that these technologies can revolutionize the theory or practice of democracy. They only see a limited role for *e*-technologies and assign three functions to them. First, the latest electronic technologies can be used to strengthen the representative institutions of liberal democracy by strengthening the link between citizens and representatives. (Wolfensberger, for instance, believes that the "democratic impulse still strongly prefers representative democracy over even a limited form of direct democracy" (Wolfensberger 2000, p.280). Second, these technologies can be employed to make it difficult for the anti-democratic tendencies to go overboard via making it difficult to control information by governments, as well as by enabling individual citizens to widen their information sources and to "utilize it against the political center" (Knei-Paz 1995, p.267). Third, these technologies can be used as the "ethos of opposition . . . to permit challenging existing hierarchies and disrupt existing constellations of power" (Schickler 1994, p.175). On the other pole of this trend lies those views which, while they acknowledge these democratic potentials of the new technologies, express the concern that they (mainly Internet) could work to the disadvantage of democracy by fragmenting and Balkanizing the public sphere (Sunstein 2001a), or by undermining the essential characters of this sphere, and subjecting the inherently slow (and desirably deliberative) features of the democratic process to the pressures from the market (Wilhelm 2000). The second trend, unlike the first, is enthusiastic about the new technologies and the promise they hold for democracy. Some of the authors belonging to this trend go as far as claiming that, at some point, the electronic media will (or should) be used as a medium for direct democracy (e.g., Grossman 1995 and Budge 1996). Slaton (1992) believes that the American political system is lagging behind the times and must be modernized by using the latest technologies to transform itself into a "Representative Participatory State." McCullagh argues that "Internet technology has the potential to reinvigorate the democratic process and re-energize citizens positively in political life" (McCullagh 2003, p.150). Similarly, Rheingold expresses the optimism that "computer-mediated communications" (CMC) could "revitalize citizen-based democracy" and could "have democratizing potential in the way that alphabets and printing presses had democratizing potentials" (Rheingold 1993, p.14, and p.279, respectively). Browning (2002) believes that the Internet can completely alter the theory and practice of advocating single-issue policies and influencing political decision-making. Other enthusiasts expect less of these technologies and imagine them to go only slightly beyond improving the representative system. Davis is of the view that the Internet would not live up to the original expectation that it could work to the benefit of small political or advocacy groups. He is also of the opinion that the Internet benefits the more traditional organizations that have larger resources (Davis 1999). Morris believes that the Internet can be used mainly as a medium to pressure the public officials to yield to the demands of the public, and that "in time, the Internet will replace the voting machine, it will become the ballot box" (Morris 1999, p.47). Moreover, some see the main democratic utilities of the latest electronic technologies lying outside of the political society and within the civil society. For instance, Becker and Wenhner argue that the Internet can only be

used to strengthen the public communication in civil society, and thus could be used in a direct manner only by "non-governmental organizations, social movements, and citizens' initiatives" (Becker and Wehner 2001, p.82). Furthermore, some authors worry about "*e*-democracy" deepening the existing inequities as they raise the problems of "digital divide," "access divide," and computer "skill divide"; or express concerns regarding privacy and security in cyberspace. (These worries and concerns will be addressed in detail in the closing section of Chapter 4.) Finally, there are those who view the foray of *e*-technologies into politics with suspicion—or regard these technologies as posing a danger to democracy or society writ large—and focus their attention on appraising the long-term social and moral aspects of the new developments as they concern themselves with studying the human condition, rapidly-changing social contexts, and the interactions between the people and *e*-technologies. For instance, Slouka (1995) argues that *e*-technologies have launched an "attack on reality as human beings have known it." All that we thought we used to be certain about, including "self," community, and "place" are under "assault" in the cyberspace. Rochlin expresses the concern that we are falling into the "computer trap" as those who design complex computerized large-scale social-technical systems "seem to have little understanding of the potential vulnerabilities they are creating" (Rochlin 1997, p.217). Wood argues that the *e*-technologies (especially radio and television) undermine our "critical thinking" abilities and the "will to reason" (Wood 2003, pp.139-40). "The more we rely on technology . . . , the less we feel we need to think for ourselves" (ibid., pp.140-41).

7. A good example here is the site <http://www.iraqbodycount.org> that offers information about the Iraqi war casualties and suffering that have resulted from the U.S. invasion and occupation of Iraq. Such information is either absent or is downplayed in the U.S. government and the mainstream media reports.

8. An excellent example here is <http://www.moveon.org>, an organization of left-leaning Democrats. Through its website, Moveon.org assists its supporters in organizing local groups or throwing "house parties" as venues for discussing issues and planning activities. In an e-mail memorandum titled "You're changing American politics" sent to its members on February 16, 2004, the organization boasted about its success in influencing mainstream politics in the United States. The starting paragraph of the memorandum made the following claim: "MoveOn is now over two million people strong in the United States. That's a huge number: the organization we've built together is bigger than the Christian Coalition at its peak." Another example here is the site <http://www.cc.org> belonging to the Christian Coalition of America which works to pull American politics in a direction opposite to the one <http://www.moveon.org> tries to pull. Browning (2002) is an enthusiastic supporter of using the Internet as a political advocacy tool and offers helpful advice to her readers concerning designing effective political websites.

9. The site <http://www.unitedforpeace.org> is a good example here, as it has shown considerable success in organizing nationwide rallies against the U.S. invasion of Iraq.

10. For example, the site <http://www.niacouncil.org> is devoted to promoting Iranian-Americans' participation in politics in the United States.

11. The site <http://www.vote.com> initiated by Dick Morris is one of these sites. See Morris (1999) for details.

12. Some of the controversies surrounding electronic voting machines and their performance in the November 2004 elections—and surrounding *e*-voting in general—will be discussed in the closing pages of Chapter 4 (pp.167-69 and notes 104 and 106).

13. The reader is advised to consult Chapter 11 of *The Betrayal of an Ideal* for a portrayal, and a critique, of the electioneering process in the late liberal democracy.

14. Some authors argue that *e*-technologies (especially the Internet) empower the traditional and well-resourced groups (e.g., Davis 1999). Others contend that well-educated and younger middle-class individuals make a wider and better use of these technologies for political purposes (Mossberger *et al* 2003). Mossberger *et al* contend that the depth of the problem of the "digital divide" is not fully understood or appreciated by researchers and policy makers. They break up the problem into three categories of "access divide," "skill divide," and "economic opportunity divide," all of which result in the "democratic divide." The questions surrounding "digital-divide" will be revisited and discussed in the closing pages of Chapter 4.

15. See Sunstein (2001a). The reader should also consult note 6 above for a short overview of some other areas of concern with the adverse political effect of *e*-technologies on society.

16. Recalling from Chapter 1, the phrase "accommodating political culture" is borrowed from Habermas (1996), p.59.

17. Rawls (1999), p.6.

18. Ellul (1992), p.48

19. The phrase "Realistic Democratic Utopia," and the fundamental idea behind it, is inspired by Rawls' "idea of realistic utopia" (Rawls 1999, p.6).

20. Some of the pitfalls and problems associated with decision-making in deliberation sessions were discussed in Chapter 13 of *The Betrayal of an Ideal*, and will be discussed further in Chapter 4 in this volume.

21. Some authors are skeptical that a popular movement for direct democracy can ever arise. Wolfensberger, for instance, is highly skeptical on this question. Notwithstanding his acknowledgement of the democratic potential of the latest electronic technologies, he believes that the "democratic impulse still strongly prefers representative democracy over even a limited form of direct democracy" (Wolfensberger 2000, p.280).

22. Recalling the caveat offered in Chapter 10 in *The Betrayal of an Ideal*, the critique of representative government offered there only targeted the *form* this conception of government has assumed in the liberal and liberal-democratic states and was not intended as an outright rejection of the *idea* itself. As will be seen later in this chapter and Chapter 4, the form of representative government adopted here has built-in features that guard against the "two democratic shortcomings" of representative government in the liberal and liberal-democratic states. See Parts II and III of *The Betrayal of an Ideal* for a discussion of these shortcomings. Chapter 11 of the work characterizes these shortcomings as "audience democracy" and "fund democracy."

23. Voting electronically, as will be seen later in this chapter and in Chapter 4, does not necessarily imply voting online. What it necessarily implies and requires is a heavy reliance on fast and powerful electronic equipment.

24. As was mentioned earlier, the percentages presented above were used mainly for illustrative purposes. The main reason for making the weights assigned to the citizens' votes directly proportional to their voting turnout rate is as follows. Lower popular voting turnouts would indicate that either most citizens do not have strong feelings or wills about the issue put to vote, or that they are content with the performance of the elected officials (the guardians and representatives), and thus trust that they would make the "right" decisions on their behalf—and this is why their votes should not be weighed heavily. Conversely, large voting turnouts would indicate that citizens have strong wills on the issue, or that they are not sure that the elected officials would make the "right" choice, and that is why they turn out in large numbers to register their wills and have

them incorporated into the decision-making process—and this is why their votes should weigh heavily.

25. As was mentioned in note 8 in Chapter 2, a voter who chooses the protest option could be offered the choice of filling out an electronic questionnaire in order to express specific reason(s) for protesting the ballot. (She would place check marks next to the statement(s) on the questionnaire that best represent(s) her reason(s) for protesting.)

26. An excellent and concise outline of the tasks the media could fulfill in a democratic political system is provided by Michael Gurevitch and Jay G. Blumler in their "Political Communication Systems and Democratic Value," in J. Lichtenberger, ed., *Democracy and the Mass Media* (Cambridge: Cambridge University Press, 1990), p.270. Habermas quotes their outline with approval (Habermas 1996, p.378).

27. The phrases "civic home" and "local talk shop" are borrowed from Barber. A civic home is a physical home of "neighborhood assembly" (Barber 1984, p.271). A local talk shop is where people socialize, such as civic homes, marketplaces, and barbershops (ibid., p.268). Moreover, the phrase "house parties" is borrowed from the literature of <http:/www.moveon.org>. See note 8 above for details.

28. Voting online does not necessarily mean voting via the Internet. As will be discussed later in this chapter and in Chapter 4, there are other options.

29. One implication of having the guardians elected at two levels is that the assemblies of the guardians could have up to 1000 or more members.

30. One advantage of Hare's method is that it allows voters to vote for more than just one candidate according to a ranking order they themselves specify. If a voter's first choice on his list gets elected, i.e., secures the "quota" (the number obtained by taking the ratio of the number of eligible voters in the national constituency to the number of the seats in the assembly), her vote will go to her second choice candidate, and so on and so forth. For a complete description, and the advocacy, of Hare's voting scheme, see Mill (1861), Chapter VII, pp.102-26, especially pp.109-10.

31. The question of property relations and property rights will be discussed in detail in Chapter 4.

32. The protective clauses in question would make the legislative process difficult for those constitutional amendments that aim at weakening direct democracy and curtailing individual rights. This could be done by requiring that such amendments be voted on twice as major issues requiring a high voter turnout rate, and a high passage rate as well. If a constitutional amendment passes, it could be required to be reaffirmed by a second vote in a year or two. The idea of protective clauses will be considered further in the following section.

33. Each of the terms "reason," "morality," and "virtue" in this context is used in a broad sense. "Reason" is used to denote expert knowledge (i.e., technical and social-scientific (and to a lesser extent "objective" or natural-scientific) knowledge, know-how, and modes of reasoning) relevant to social-political decision-making. Moreover, it is also used to denote theoretical knowledge that understands the deeper social-philosophical issues and takes a long-term view of society and its aims and interests. The term "morality," on the other hand, stands for the importance of normative considerations or moral thinking in political decision-making, vis-à-vis the sort of influence the considerations of economic utility, efficiency, and monetary interests (often favoring special-monetary interests) can exert in social decision-making. Finally, "virtue" here underscores the importance of the need for truth, integrity, and sincerity in politics and political office. As will be discussed further in Chapter 4, the guardians are conceptualized as representing both the social-moral conscience and the ideally high social-political consciousness of the

citizens. They are conceptualized as being well-versed in matters of social, political, and economic policy and philosophy, either by their superb education and research, or by their exemplary experiences in serving their communities, as well as being judged by the citizens to possess these qualities and as being trustworthy and having strong moral character. The guardians are also conceptualized as representing the citizens' aspirations for embedding social-political decision-making in knowledge, truth, and virtue, as well as being individuals who are perceived as leading their own lives in accordance with these principles. They bring knowledge, truth, reason, and morality to politics, not only by directly partaking in decision-making in their respected assemblies, but also by influencing and informing the citizens' will-formation activities. They do so by actively engaging the citizens dialogically in deliberations and discussions, and also by being sources of knowledge and social-political wisdom, to which citizens can turn and consult in the process of forming their public wills.

34. The guardians of the Realistic Democratic Utopia would regard interests as "objective," "broad," and of "one nation . . . of the whole," unlike the representatives who would take them as "attached," "personal," "subjective," and conflict-ridden (as presently is the case in the U.S. Congress). The discussion in Part II of *The Betrayal of an Ideal* moved in the direction of regarding the first set of the attributes of interests as "Burkean," and the other as "Madisonian."

35. This manner of balancing the power of the guardians, vis-à-vis that of the representatives, was alluded to in Principle III. The guiding principle for this balancing of power is based on the following two presuppositions. First, large voting turnouts, in addition to the reasons stated in note 24 above, could also be attributed to the success of the guardians in encouraging citizens to participate in voting. Therefore, assigning a larger weight to the votes of the guardians, vis-à-vis those of the representatives, can be regarded as a "reward" system for encouraging the guardians to continue to engage the public in politics. The more they draw the public to participate in major issue voting occasions, the more their own votes count. Second, on the other hand, large voting turnouts can also be attributed to a lack of trust of the citizens in the elected officials. This would be an indication that society is at the threshold of a critical stage, and thus it could use more good judgement than it would need in ordinary circumstances; and given that the guardians have higher credentials than the representatives in this respect (see Principle VII), their votes should carry more weight than those of the representatives in these circumstances.

36. As was mentioned earlier, the notion of voting electronically does not necessarily imply voting on-line or voting from home. However, as will be seen later in this chapter and in Chapter 4, the notion does necessarily imply and require a heavy reliance on fast and powerful electronic equipment.

37. One should recall that Chapter 2 discussed some other options for choosing the winner in the absence of a clear favorite in the primary, some of which would make it possible to identify the winner the same day.

38. See Principle VI above for a possible set of requirements.

39. This claim is adequately substantiated in Parts I and II of *The Betrayal of an Ideal.*

40. The role of civic associations will be discussed in greater detail in Chapter 4. Here, it suffices to add that civic associations, despite the fact that they often have single-issue orientations (e.g., religious associations, civic groups, race- and ethnically-based groups, consumer interest groups, environmental associations, gay and lesbian interest groups), as Iris Marion Young has argued, do not necessarily and primarily work for the "collective self-interest of their [own] members" (Young 1995, p.209). As will be discussed in Chap-

ter 4, these associations often coalesce with one another; and the issues they advance often echo, or have bearing on, the concerns of other associations; and consequently, the issues they raise impact the society as a whole.

41. As a minimum program, one can remove the assemblies of the guardians (the trustees and experts) altogether. This would mean that the guardians would be excluded from voting or taking part in decision-making. Their function then would be limited to facilitating the political education of the public and drafting the laws and sending them to legislators and the public for approval. (Their function as the drafters of the laws and policies make them similar to J. S. Mill's special commission of the experts.) Now, in this minimum program, if the guardians cannot partake in legislating (i.e., voting side by side with the representatives and the public), then there seems to be no point in having them elected by the larger public. In the minimum program, the guardians can be chosen by lot from a pool of candidates proposed or recommended by the civic associations, universities, non-profit organizations, and other institutions in the "civil society." (The pool would consist of many sub-pools divided into the categories of philosophers, economists, social scientists, trustees, etc.) As was the case with the original maximum program above, the guardians will have no links to the "economic society" in the minimum program.

42. A reference to the famous utterance of Oscar Wilde. Cf. Levine (1993), p.92.

43. Rousseau (1762), *The Social Contract*, Book III, p.74.

44. This point was discussed in some detail in Chapter 13 of *The Betrayal of an Ideal*.

45. Budge (1996), p.25. Iain McLean estimates the number of citizens in Athens just over 40,000 (McLean 1989, p.5). According to McLean, although the size of the assembly amphitheater is 6,000, "far fewer than 6,000 people turned up for ordinary sessions" (ibid., p.6).

46. "Perhaps an ideal deliberative procedure is best institutionalized by ensuring well-conducted political debate among elites, thus enabling people to make informed choices among them and the views they represent" (Cohen 1997b, p.422).

47. Historically, though parties have appeared as putting forth platforms for the general interests of society, broadly speaking, they have advanced mainly special economic or class interests. One major reason for the decline of party loyalty has been the fading away of the economic opposition (the left versus the right) that was a determining factor in the early-to-mid-twentieth century. (This opposition lasted up to the late 1980s and ended with the victory of the right.) Another major reason has to do with the rise of the "new social movements" which have provided new resources for citizens in the way of grand conceptions of the common good (e.g., the ecological movement, peace movement, and women's movement). Among other reasons, one should mention the rise of interest in "single-issue" politics and the ever-rising ethnic diversity of citizens—(new immigrants and their immediate offspring often do not identify with the parties of their new country).

48. Manin discusses some aspects of the rise of independent thinking among the party members (Manin 1997, pp.203-35, e.g., see p.219 and p.231). "The number of floating voters who do not cast their ballot on the basis of stable party identification is increasing. A growing segment of the electorate tends to vote according to the stakes and issues of each election" (ibid., p.231). Moreover, "[v]oters tend increasingly to vote for a person and no longer for a party or a platform" (ibid., p.219).

49. Despite the weakening of party loyalty as manifested in the voting patterns of voters, political parties will continue to remain indispensable to the political process in the Realistic Democratic Utopia. For a discussion of the importance of the role of parties in a

direct democracy, see Budge (1996), e.g., p.175; also see Budge's "Deliberative Democracy Versus Direct Democracy—Plus Political Parties" in Saward (2000), especially pp.206-09.

50. As was discussed in Part II of *The Betrayal of an Ideal*, the major political parties in the liberal and liberal-democratic states (especially in England, the United States, and Canada) by and large represented the class interests of the major propertied classes or the loosely organized sectional and regional upper- and middle-classes. However, starting with the process of transition to the liberal-democratic state, these parties began to position themselves such that they could be projected as representing the common interests of all. See Macpherson (1977), pp.64-69, for a discussion of this "blurring [of] class lines" by political parties and the role played by this "blurring" in moderating the class conflicts in transition from the liberal state to the liberal-democratic one.

51. The sense of "social capital" used here is the one defined by Putnam: "the core idea of social capital theory is that social networks have value . . . social capital refers to connections among individuals—social networks and the norms of reciprocity and trustworthiness that arise from them. In that sense social capital is closely related to what some have called 'civic virtue.' The difference is that 'social capital' calls attention to the fact that civic virtue is most powerful when embedded in a dense network of reciprocal social relations. A society of many virtuous but isolated individuals is not necessarily rich in social capital" (Putnam 2000, pp.18-19).

52. Here, it is taken for granted that the representatives in the Realistic Democratic Utopia will behave in similar ways as their predecessors have done historically, that is to say, they will be susceptible to the influence-peddling and lobbying of the economic society. This discussion will be taken up further in Chapter 4.

53. This could be done by stipulating difficult procedures for those amendments to the constitution that would be devolutionary in character and would have dismantling effects on the system of the direct democracy of the Realistic Democratic Utopia. The protective clauses in question could require that amending bills be voted on twice as major issues, requiring a high voter turnout rate, and a high passage percentage rate as well. If a bill amending the constitution passes in the first round of voting, it would need to be reaffirmed by a second vote in a year or two. An amendment will not be adopted if it fails to be reaffirmed in a second vote.

54. In essence, this is Madison's famous argument in the "Federalist No. 10." "Extend the sphere, and you make it less probable that a majority of the whole will have a common motive to invade the rights of the other citizens; or if such a common motive exists, it will be more difficult for all who feel it to discover their own strength, and to act in unison with each other" (Hamilton 1961, p.48).

55. McLean (1989), pp. 61-107. CATI (Computer-Assisted Telephone Interviewing), a cable-TV-based electronic voting system called "Qube," and "Consensor," an electronic voting system that registers the intensity and weight of the voters' vote, are among these systems and technologies.

56. Budge (1996), p.27. Budge's survey of literature on the feasibility of using modern technology for expanding democracy is instructive.

57. Wolff (1970), pp.34-35. Wolff's system of "instant direct democracy" would function in the following way: "Each evening, at the time which is now devoted to news programs, there would be a nationwide all-stations show devoted to debate on the issues before the nation. Whatever bills were 'before the Congress' . . . would be debated by representatives of alternative points of view. There would be background briefings on technically complex questions, as well as formal debates, question periods, and so forth.

Committees of experts would be commissioned to gather data, make recommendations for new measures, and do the work of drafting legislation. One could institute the position of Public Dissenter in order to guarantee that dissident and unusual points of view were heard. Each Friday, after a week of debate and discussion, a voting session would be held. The measures would be put to the public, one by one, and the nation would record its social preferences instantaneously by means of the machines. . . . Simple majority rule would prevail, as is now the case in the Congress."

58. Ibid.

59. The reader may want to glance over note 1in the Introduction to this volume for a short statement on the moral content of the original idea of democracy.

60. See Chapter 1 for a discussion of this point.

61. The literature on the questions of democracy at the workplace and the local community is rapidly growing. The following are some examples of titles on the topic of the democratic community: *Community* (Delanty 2003), *Deep Democracy: Community, Diversity, and Transformation* (Green 1999), *Civic Virtues: Rights, Citizenship, and Republican Liberalism* (Dagger 1997), *Who Cares? Rediscovering Community* (Schwartz 1997), *Genuine Individuals and Genuine Communities* (Kegely 1996), *The Search for Political Community: American Activists Reinventing Commitment* (Lichterman 1996), *New Communitarian Thinking: Persons, Virtues, Institutions, and Communities* (Etzioni 1995), *Liberalism and Community* (Kautz 1991), *Human Rights and Search for Community* (Howard 1995), *Democratic Community* (Chapman and Shapiro 1993), *The Moral Commonwealth: Social Theory and the Promise of Community* (Selznick 1992), *The Dance with Community* (Flower 1991), *The Idea of a Democratic Community* (Berry 1989), *Strong Democracy* (Barber 1984), and *Reclaim the State* (Wainwright 2003). The reader should consult note 74 in Chapter 1 for a list of titles on the topic of the workplace democracy.

Chapter 4

The Theory of Direct-Deliberative *e*-Democracy

This chapter brings together the "conception of democracy as the political empowerment of the citizen," the scheme of amalgamating-composing collective wills, and the theoretical framework of the "Realistic Democratic Utopia" developed in the preceding chapters, and synthesizes them into a single theory of democracy that will be referred to from this point on as the "*theory of direct-deliberative e-democracy*."[1] As will be seen, this synthesis takes place against the backdrop of the "liberal-democratic conception of democracy," on the one hand, and the theories of participatory and deliberative democracy, on the other hand. The theory of direct-deliberative *e*-democracy has two main components. The first is a conception of the social and political empowerment of the citizen developed in Chapter 1. The second is a theoretical justification for a set of institutional arrangements and collective decision-making schemes that offer a practical expression to this empowerment. Explored in Chapters 2 and 3, this second component necessarily ventures into the realm of seeking technical solutions to some of the theoretical difficulties that have dogged democratic theory throughout its long history.

The main political value affirmed by the theory of direct-deliberative *e*-democracy is the value inherent in the principle of the citizens' direct participation in legislating the major laws that shape or affect their lives. On this account, the theory is indebted to Rousseau. The second political value affirmed by the theory of direct-deliberative *e*-democracy is the principle that social-political decision-making ought to be embedded in knowledge, moral understanding, and virtue—and on this, the theory owes much to Plato's theory of political guardianship. The Rousseauean connections of the idea of the citizens' direct participation in collective decision-making were explored in Chapter 1. The Platonic connections of the theory, on the other hand, were hinted at in Chapter 3 and will be explored further in this chapter and the Conclusion.

Putting aside for a moment the question of the philosophical underpinnings of the theory of direct-deliberative *e*-democracy, what needs to be stressed at the outset is that the development of the theory itself is motivated by the latest innovations in *e*-technologies. These innovations have given rise to an intellectual-political movement that is committed to exploring the ways in which these technologies can be used for furthering the cause of democracy in present-day American society. Some of the theories and ideas that have come out of this movement were surveyed in Chapter 3. The theory of direct-deliberative *e*-democracy belongs to this movement. The main goal of the theory is to place the building-blocks of the fundamental ideas that have motivated these theories on philosophical grounds, and attempt to develop a philosophical framework for the idea of *direct e-democracy*. The philosophical foundation of this framework was laid down in Chapter 1. Starting with this foundation, and using the explorations of Chapters 2 and 3, the present chapter will attempt to complete the framework by elaborating the theory of direct-deliberative *e*-democracy. To this end, in what follows, first an outline of the essential features of the theory will be presented. This presentation will then be followed by a number of discussions in which the theory will be compared with other theories of democracy. The remaining features of the theory will unfold in the course of these discussions.

As to its most essential features, the theory of direct-deliberative *e*-democracy, first and foremost, is committed to the realization of the idea of the unencumbered and fullest feasible (positive) development of the human individual, which it regards as being the ultimate value of the human universe and society's *raison d'être*. Moreover, the theory is built around a faith in the abilities of ordinary citizens to make sound political decisions. (As will be seen later in this chapter, this faith is a qualified one.) Furthermore, the hallmark of the theory of direct-deliberative *e*-democracy is that it rests on a thick notion of sovereignty. The theory is relentless in the pursuit of the idea that democracy is to be primarily identified with the citizens' *direct* and *continuous* exercise of sovereignty. On this question, the theory diverges considerably from the liberal-democratic conception that is premised on a limited, indirect, and intermittent-periodic exercise of sovereignty by citizens that takes place exclusively during the election of the independent-virtual representatives. Another feature of the theory of direct-deliberative *e*-democracy is that its thick notion of sovereignty is coupled with a thin notion of equality in the realm of material holdings. This thin egalitarianism is a direct consequence of developing the theory within a liberal-democratic framework. As will be discussed in some detail later in this chapter, the thin egalitarianism of the theory goes against the grain of the tradition in the "classical theory" of democracy, and hence needs to be justified, and defended against potential criticisms.[2] (This will be done in the third section of the chapter.) For the time being, it should suffice to note that the conception of equality espoused by the theory of direct-deliberative *e*-democracy is much thicker than the one present in the liberal-democratic conception, and that it appears thin only if it is viewed from the perspective of the "classical theory."

Finally, the other essential feature of the theory of direct-deliberative *e*-democracy is its own particular notion of democratic legitimacy. In sharp contrast to other theories of democracy, the theory of direct-deliberative *e*-democracy is strongly committed to the view that the question of democratic legitimacy is—not ultimately, but *immediately* and *directly*—the question of the individual citizens' sovereignty. The question of democracy or democratic legitimacy is not primarily about giving to people the "freedom of choice" in politics or the "right to choose" their governments (liberal democracy); nor just about securing their consent (deliberative democracy, liberal-democratic conception); nor just about establishing "procedures" (liberal-democratic conception) and assuring their "fairness" (deliberative democracy), nor just about morally justifying the power of authority and its right to exercise this power (liberal-democratic conception, deliberative democracy). Nor is it just about the righteousness of the "intent" or the ultimate ends of the policies and the laws that authority imposes and legislates (proletarian democracy). Neither is this legitimacy just about the "outputs" and effects of the laws legislated and the policies instituted—i.e., the "content of outcomes" or "substance" of decisions made by the authority—(participatory democracy, deliberative democracy, proletarian democracy); but also and primarily about the *actual, direct, and continuous input* of individual citizens into the social and political decision-making process. *More than anything, democracy is about individual citizens experiencing the political power directly and doing so on a continuous basis.* Consequently, the yardstick of democratic legitimacy is the degree to which this ideal is realized. Democracy is about providing and facilitating the highest feasible degree of the actual, i.e., *direct* and *ongoing*, participation of citizens in legislating the fundamental laws they abide by, and in making decisions about the fundamental policies that affect their lives. For the theory of direct-deliberative *e*-democracy, this constitutes both the immediate and ultimate yardstick of democratic legitimacy.

One should note that this notion of democratic legitimacy is not about granting to people what they will, nor about ruling them in accordance with their wills. Rather, it is about *empowering the people to do the ruling themselves*—this constitutes the primary focus of the theory of direct-deliberative *e*-democracy. This idea of democratic legitimacy takes for granted the availability of a social decision-making process that grants to citizens two sorts of power: the power to fully express their political wills, and the power to have these expressions fully integrated into a procedure that they generally regard as a reasonably fair and accurate mechanism for amalgamating their individual wills and composing collective wills out of them. Moreover, this proceduralist legitimacy is further enhanced by the outcome-based "moral-epistemic legitimacy" of the collective wills produced in the process. This latter legitimacy is drawn from the availability and the proper functioning of the public institutions of civic education and deliberation. That is to say, on the strength of these institutions, the collective wills produced in direct-deliberative *e*-democracy would also lay claim to being the embodiment of *both* the wisdom and public judgements of a civic-minded and well-educated body politic *and* the reasoned judgements of a well-informed and knowledgeable public—

and thus to being morally-informed and fairly-rendered collective judgements, as well as to being epistemically reliable.

These preliminaries lay down the foundations of the theory of direct-deliberative *e*-democracy in a summary form. The full elaboration of the theory will take place in the following pages, where the theory will be fully developed within the context of comparing it with other theories of democracy. Moreover, the numerous discussions that will address questions of private property, technology, guardianship, and civil society will also contribute to the goal of elaborating the theory.

Direct-Deliberative *e*-Democracy and Liberal Democracy

As has been emphasized repeatedly, the theory of direct-deliberative *e*-democracy is developed against the backdrop of the present-day American liberal democracy. Far from being the negation of liberal democracy, the theory of direct-deliberative *e*-democracy is squarely grounded within a theoretical framework whose underlying ideas are present in the existing "public political culture" of the American society.[3] This is evident in the emphasis the "conception of democracy as the political empowerment of the citizen" places on the need for individual rights and liberties, and on the equal access of all citizens to them. In this respect, the theory of direct-deliberative *e*-democracy is in conformity with the "democratic consciousness" of present-day society. Direct-deliberative *e*-democracy is a *constitutional democracy*. Its constitution lays down a set of "first and higher laws" that guarantees a rich package of universal and equal basic rights and civil liberties and, at the same time, puts these rights and liberties outside of the domain of democratic decision-making.[4] The constitution of direct-deliberative *e*-democracy also contains a set of clauses that impede attempts to weaken these rights and liberties or to amend the constitution in ways that could endanger these rights or liberties, or facilitate their erosion. In other words, the constitution of direct-deliberative *e*-democracy has built-in features that stand in the way of voting democratically to undermine its liberal components.[5]

The theory of direct-deliberative *e*-democracy starts with a rich liberal-democratic conception of the individual's rights and (negative) liberties and couples it with its own democratic conceptions of the individual and society. The latter conceptions, though they are not characteristically liberal-democratic, nonetheless have strong liberal-democratic features. More specifically, the conception of the individual developed in Chapter 1 starts with the premise that Man is "essentially" a social being, and then integrates into itself the essential features of the conception of the individual in liberalism. The conception of society espoused by the theory of direct-deliberative *e*-democracy, on the other hand, is in many respects similar to Rawls' conception of society as a cooperation of free and equal individuals. The theory of direct-deliberative *e*-democracy draws these conceptions together and fuses them within a framework that permits the introduction of the "conception of democracy as the political empowerment of the citizen." In this respect, the theory of direct-deliberative *e*-democracy, one can argue, is a theory of the actualization of the liberties that liberal democracy bestows on the individual, and the political powers that it attributes to her, without empowering her to exercise them. This contention will be elaborated in what follows next.

In one sense, the main purpose of the theory of direct-deliberative *e*-democracy is to reconcile, and conjoin, the importance liberalism assigns to negative liberties and the privacy of the individual, on the one hand, with the emphasis the original idea of democracy places on the citizen's direct participation in political decision-making, on the other hand. In direct-deliberative *e*-democracy, this reconciliation takes place at the level of the "individual citizen" and within a framework that seeks a synthesis of theory and practice.

Now, on the question of the individual's direct participation in democracy, the theory of direct-deliberative *e*-democracy parts company with liberalism and stands opposed to it. Liberalism, despite its unswerving commitment to the sanctity of the individual's liberties and privacy, and despite its vehement defense of these liberties against the public's intrusion, lacks faith in the individual's intellectual and moral competence in attending to matters of the public interest—(albeit it takes the individual as the ultimate moral-intellectual authority on matters of her self-interests or self-seeking pursuits). In this respect, liberalism, one can argue, is committed to an apolitical-asocial and passivist brand of individualism. The theory of direct-deliberative *e*-democracy, by contrast, puts forth an activist-capable notion of individualism that postulates the individual as socially interested, and willing to participate in the public's affairs, as well as postulating her as possessing the intellectual-moral capabilities for doing so. Recalling from Chapter 1, the individual in the theory of direct-deliberative *e*-democracy is conceptualized as a fluid and dynamic unity of private and public polarity as well as the self-interested and socially-oriented polarity. Moreover, in addition to being founded on the philosophical conception of the individual developed in Chapter 1, the activist-capable brand of individualism espoused by the theory of direct-deliberative *e*-democracy is also premised on a number of social-political assumptions and contentions that can be categorized into the following presuppositions:

1. The knowledge of the "truth" and virtue (morality), and political know-how, and the ability to reason are not the exclusive property of intellectuals and political elites. An individual who is sufficiently informed and educated in the public issues and civic virtues, responsibilities, and skills, who lives in a social environment and culture that is conducive to free and democratic thinking and behaving, is capable of contributing to the making of morally sound and politically reasonable decisions in matters of the common good.

2. Individuals who are sufficiently informed and educated in the public issues and civic virtues, responsibilities, and skills, who live in a social environment and culture that is conducive to free

and democratic thinking and behaving, are the most qualified judges and defenders of their own interests.[6]

3. If given the opportunity, an individual who is sufficiently informed and educated in the public issues and civic values, responsibilities, and skills, who lives in a social environment and culture that is conducive to free and democratic thinking and behaving, would be willing to participate in decision-making in matters of the common good.

4. Disagreements about societal issues and policy-making matters among citizens often arise from the mutual ignorance of the views and interests of other individuals or contending groups, and thus, are reconcilable through deliberation and negotiation.[7]

Rather than being seen as fanciful beliefs of an idealist democrat, one can argue that these propositions should be regarded as optimistic interpretations of the empirical observations about the social-political behaviors of the citizens in present-day America, especially in the period of the last four decades or so. These propositions complement the activist-capable conception of the individual developed on a philosophical plane in Chapter 1. These contentions will be revisited later in this chapter in arguing against the pessimistic appraisals of the ordinary citizens' moral and political aptitudes, as well as in rejecting arguments against the citizens' direct participation in social decision-making.

Another advantage of the theory of direct-deliberative *e*-democracy over liberal democracy is that the former postulates democracy as a *positive conception*, in which the negative liberties of the individual (especially her freedom of expression) have been extended to the *positive* realm of actualization. In other words, direct-deliberative *e*-democracy gives a positive content and a real force to the negative liberties espoused by liberalism (especially the freedom of expression) and by doing so bridges the gap between the negative liberties and "positive liberties."[8]

Furthermore, the characterization of direct-deliberative *e*-democracy as a *positive conception* of democracy is also intended to distance direct-deliberative *e*-democracy from two *negative* characterizations that have clouded the idea of democracy throughout its history. The first characterization in question is relatively new and is associated with how liberal democracy conceptualizes citizens' participation in the democratic process. In liberal democracy, the citizens' participation in politics is *negative* in that citizens affect or "control" policy-making and the legislative process (and hence control or curb the tendency toward tyranny or corruption) by the threat of having the voting power to unseat elected officials, if they fail to live up to the public's expectations.[9] Of course, the theory of direct-deliberative *e*-democracy can also be characterized negatively in this sense. This is because direct-deliberative *e*-democracy, in addition to being a direct-participatory democracy, is also a strong and vibrant *representative democracy*. As a matter of fact, direct-deliberative *e*-democracy has larger, and much more potent, representative bodies and structures than liberal democracy, and thus the negative control of the officials is of utmost importance in direct-deliberative *e*-democracy. Notwithstanding the importance of this negative participation in direct-deliberative *e*-democracy, it is worth reiterating that the main aspect of the citizens' participation in direct-deliberative *e*-democracy is *positive*. In direct-deliberative *e*-democracy, while the negative sense of the citizens' participation functions in the same way as it does in liberal-democracy, in its positive sense, participation in direct-deliberative *e*-democracy corresponds to the third concept of positive liberties defined by Macpherson.

Concisely stated, in unpacking Berlin's concept of positive liberties, Macpherson breaks up the notion into three *different* concepts of "individual self-direction" (or "developmental liberty"), "coercion," and "participation"(or "liberty as a share in the controlling of authority").[10] No doubt, the second concept—(i.e., "liberty . . . [as] . . . coercion . . . by the fully rational or by those who have attained self-mastery . . . [or] . . . by those who say they know the truth, of all those who do not (yet) know it")—poses a problem for the idea of participatory democracy and must emphatically be rejected by its proponents as part and parcel of the effort to retrieve the true content of democracy.[11] Macpherson is of the view that this second sense of positive liberties is the "idealist or metaphysical rationalist transformation" of the first sense and thus is "unfortunate" and must be cast aside.[12] Now, given that the theory of direct-deliberative *e*-democracy is emphatic concerning the protection of negative liberties, it clearly cannot espouse any notion of positive liberties that could remotely interpret liberties as "coercion." Notwithstanding this strong position against the second concept, as will be argued later in this chapter, the positive consequences that have often been assumed to spring from "coercion," and thus have been used as justifications for it—(viz., the need for truth, knowledge, reason, morality, and virtue in governing)—are absorbed by the theory of direct-deliberative *e*-democracy in its conception of guardianship. Via this concept, as will be seen below, the notion of "coercion" will give way to the idea of "suggestion through education"; and the potential subjugation of "those who do not (yet) know" by "those who say they know the truth" will be transformed into a system of *political partnership* and *power-sharing* of these two seemingly adversarial groups.

The second negative characterization that has clouded the idea of democracy is the criticism that democracy is an expression of the reaction of many, i.e., the ordinary, the weak and have-nots, against the few, i.e., the successful, and the haves. Democracy, as this criticism goes, is a manifestation of the desire of the many to drag down the few (the successful) to their own low and common levels. In this negative characterization, democracy is portrayed as an expression of reaction against progress and advancement, and the embracement of mediocrity and backwardness. Without a doubt, this is an elitist criticism that targets the egalitarian component of the classical theory of democracy. This criticism originated in ancient Athens and has had a ubiquitous presence since then.[13] In the modern world, Nietzsche is a good representative of this negative view on democracy.[14] Liberal democracy escapes this negative characterization by offering only a "minimal" democracy, i.e., by fully embracing the first negative characterization that keeps citizens (the many, the ordinary) at a safe distance from the business of governing. The theory of direct-deliberative *e*-democracy also escapes this second negative characterization; not by embracing the other negative characterization as liberal democracy does, but rather through attempting to mitigate the prevailing dichotomous opposition between citizens, on the one hand, and the elites, on the other hand. In doing so, the theory does not attempt to do away with elites or to disempower them. Rather, it does so by assigning to elites a greater importance than the one presently assigned by liberal democracy. That is to say, direct-deliberative *e*-democracy not only employs elites as the independent-virtual representatives of the people (as is the case in liberal democracy) but also engages them dialogically with the rest of the citizens as their social-political "mentors" or "educators." In this capacity, i.e., as guardians of the public interests, elites work toward elevating citizens to higher levels of social-political knowledge and moral competence.

As the final comparison between the theory of direct-deliberative *e*-democracy and liberal democracy, one can return to the main themes of Parts II and III of *The Betrayal of an Ideal* and compare the way these theories utilize the representative form of government. The main criticisms directed against the liberal-democratic conception in *The Betrayal of an Ideal* are the contentions that the form in which the liberal-democratic conception of democracy employs representative government has two main democratic shortcomings, and that the liberal-democratic conception itself amounts to the political disempowerment of citizens.[15] The specific form of representative government adopted by the theory of direct-deliberative *e*-democracy in Chapter 3 confronts directly, and deliberately, these shortcomings of the liberal-democratic form. It is in this light that the theory of direct-deliberative *e*-democracy poses itself as the antithesis to the democratic shortcomings of the liberal-democratic conception of democracy.

Direct-Deliberative *e*-Democracy and Theories of Participatory and Deliberative Democracy

Though the theory of direct-deliberative *e*-democracy has a solid and vibrant component of representation, in light of the fact that the citizens' participation in decision-making is direct, it is essentially a theory of *constitutional direct democracy*. Moreover, given that the social process of policy and legislative decision-making in the direct-deliberative *e*-democracy is conceptualized as being essentially a will-formation activity that would commence only after the completion of a period of extensive public discussions and deliberations, the theory of direct-deliberative *e*-democracy can also lay claim to having a strong deliberative component. In light of these considerations, it would be more appropriate to characterize the theory of direct-deliberative *e*-democracy as a theory of *constitutional direct-deliberative democracy*. Finally, given that the public discussions-deliberations and the actual process of socially composing the collective wills would involve the participation of both individual citizens and elites (i.e., the guardians and virtual representatives), the theory of direct-deliberative *e*-democracy can also be characterized as a theory of *power-sharing of the elites and ordinary citizens*.

One advantage of the theory of direct-deliberative *e*-democracy over the theories of participatory and deliberative democracy is that, in contrast to latter theories, the theory of direct-deliberative *e*-democracy gives a concrete and direct expression to the ideal of citizens' participation and deliberation without limiting, or appearing to limit, the scope and substance of individual rights and liberties. As was discussed in Part III of *The Betrayal of an Ideal*, this was an important area of difficulty for the theories of participatory and deliberative democracy, especially for the former. Another advantage of the theory of direct-deliberative *e*-democracy over these theories is that, in contrast to their reluctance to address the question of model making and institutional arrangements necessary for the realization of the ideas of the citizens' deliberation and participation, direct-deliberative *e*-democracy faces the question head-on. In the case of the theories of deliberative democracy, the boldest attempts in this direction stop at arguing in favor of notions such as "distributive measures" and "public funding of the political parties."[16]

In discussing the main differences between direct-deliberative *e*-democracy and the theories of deliberative democracy on the questions of conceptualizing the idea of democracy and its actual workings, one should point to the over-emphasizing of the notion of "reason" in latter theories.[17] For instance, in the case of Habermas, democracy is conceptualized instrumentally as a "procedure" in the service of "rational will formation."[18] One consequence of over-emphasizing reason in these theories is that they are led to overburden the individual rationally. In direct contrast to liberalism, which takes the human individual as an isolated being in society and regards her political behaviors (e.g., her contribution to social decision-making) as expressions of her self-interests, deliberative democracy falls into the other extreme of expecting the individual to cast aside her self-interests and join the collective decision-making process as the bearer of the common good. While liberalism takes self-interest as the main, if not the only, guide to the individual's social-political behavior and thinking, the earlier theories of deliberative democracy expect reason to be her main, if not her only, guide in her social-political activities. In contrast to deliberative theories, the theory of direct-deliberative *e*-democracy does not overburden the individual by demands of rationality or morality.[19] Rather, it only attempts to employ reason and morality in the service of the *democratic* will-formation. The theory of direct-deliberative *e*-democracy does not require the individual to suppress her self-interests, or surrender her "private wills" that represent these interests to her "public wills" in the moral plane; neither does it demand that the common good ought to be the only relevant criterion in the rationally or morally based decision-making on the public issues.[20]

Another major difference between direct-deliberative *e*-democracy and the theories of deliberative democracy has to do with the actual workings of democracy, i.e., with the process of collective will-formation, as well as with the procedures for composing the collective wills. Insofar as these questions are concerned, the theory of direct-deliberative *e*-democracy conceptualizes democracy as a social practice that *both transforms* the wills of citizens, be they private or public, *and amalgamates-composes* a social decision (or a collective will) out of them. The *transformative* aspects of direct-deliberative *e*-democracy has to do, on the one hand, with what Nino calls the "moralization of people's preferences" and, on the other hand, with raising their level of factual and theoretical knowledge on the issues—both of which would take place in the process of public deliberation *and* under the guidance and leadership of the guardians.[21] On the question of actualizing the conceptualization of democracy as a social practice of *amalgamating-composing* collective wills, the theory of direct-deliberative *e*-democracy turns to the latest *e*-technologies, and the complex amalgamation-composition schemes they bring about, in order to attain some reasonable and acceptable levels of "fairness" and "accuracy."

Now, given that politics is not just the science of realizing the good of the community by doing morally right things or acting in accordance with the dictates of reason and knowledge, but also, the art and the practical prudence of doing the reasonable thing, as well as attending to the everyday needs and interests of the citizens, it then follows that the social practice of decision-making must incorporate *all* of these considerations, including the private-interests of the citizens. Moreover, in light of the second presupposition of the activist-capable conception of individualism stated above, it also follows that citizens ought to express the wills that represent their interests (both public and private) *directly,* in addition to having them expressed by their elected officials, i.e., the guardians and independent-virtual representatives.[22] In these two respects, the theory of direct-deliberative *e*-democracy stands opposed to the theories of participatory and deliberative democracy.

It should be noted at this point that although the electronic scheme for composing the collective wills in direct-deliberative *e*-democracy is proposed primarily for macro levels, there are no good reasons to believe that this scheme would be of no use at micro and meso levels. Moreover, it should be emphasized that, while the transformative aspect of the scheme at macro levels (and perhaps at meso levels as well) is understood in terms of individuals educating themselves under the directorship of the guardians, at micro and meso levels, this aspect would take deliberative and dialogical character. Having emphasized this, one should add that, unlike what the theories of deliberative democracy aim at, the goal of education and deliberation at these three levels, especially at macro levels, is not to reach a *consensus*, but rather to develop a common understanding of the issues—as much as possible—and perhaps reach a *compromise* in the case of the micro levels.[23] No doubt, achieving consensus would be ideal. However, requiring it could be impractical at meso levels and much harder, if not impossible, at macro levels. Such a requirement could also be oppressive and strain the individuals' liberties at the micro levels, in light of the likelihood that, under the pressure of seeking consensus, some individuals would express wills other than their true ones when the time arrives to reach a decision. Analogous to the inclination of the liberal-democratic conception to regard the occasion of casting votes as a private matter, the theory of direct-deliberative *e*-democracy is insistent that decision-making ought not to take place in the deliberative space and time, but later in the individual's own *private space and moment*. This is mainly intended to provide the individual with time to reflect upon the issue to be decided, consider the arguments presented, and make the cognitive and moral transition from holding a mere opinion on the issue, and having a mere preference for a policy option, to having formulated her wills on the issue—whether public or private.

Now, despite its insistence on the "sanctity" of the individual casting her vote in private space and moment, the theory of direct-deliberative *e*-democracy is committed to the idea of *politics as dialogue* to the same extent that the theories of deliberative democracy are. Compared with the theories of deliberative democracy, although direct-deliberative *e*-democracy facilitates and expands the public channels for the deliberation and political education of the citizenry, when it comes to the actual moment of arriving at social decisions, it allies itself with liberal democracy, as it relies solely on the wills of citizens expressed in private.[24]

An alternate approach to highlighting the differences between the theory of direct-deliberative *e*-democracy, on the one hand, and the theories of deliberative democracy, on the other hand, would be to discuss how these theories address the question of the citizens' input into social decision-making. With Joshua Cohen's version in mind as a case in point, one can conceptualize deliberative democracy as a framework or procedure for transforming the "expressive liberty" of the citizens (provided and protected in liberal democracy) into what one might call its "deliberative empowerment"—that is to say, a framework for giving the citizens' deliberation some role in influencing the decision-making process.[25] The theory of direct-deliberative *e*-democracy, in comparison, goes beyond this limited framework. It takes the expressive liberty and transforms it into "expressive empowerment," as well as into "legislative empowerment"—that is to say, it empowers citizens to *directly* express their wills and *directly* transform their expressed wills into social decision-making powers.

Despite these differences between the theory of direct-deliberative *e*-democracy and the theories of deliberative democracy, it is important to emphasize that direct-deliberative *e*-democracy does not negate deliberative democracy, nor does it counterpoise itself to it, or compete with it. On the contrary, given that its main target arena is the macro levels, direct-deliberative *e*-democracy has the theoretical space and flexibility to integrate theories of deliberative democracy into itself as complementary theories which address the question of democracy at micro and meso levels. For instance, in the case of Habermas' conception, deliberation at the level of associations is understood in terms of engagement in public opinion- and will-formation activities.[26] In this capacity, deliberative democracy can be regarded as complementing the theory of direct-deliberative *e*-democracy directly.

Deliberative democracy can also complement direct-deliberative *e*-democracy indirectly, in the Habermas' conception, through using the associations as forums for nominating the candidates for guardianship positions, as well as using them as vehicles for campaigning for their election into these positions. These two levels and forms of activity in associations reflect some aspects of the current practices of associations in American liberal democracy. Associations such as the National Association for the Advancement of Colored People (NAACP) and the National Organization for Women (NOW) are two cases in point here. Moreover, it is relevant here to add that the associations in question, despite the fact that they often have single-issue orientation, (e.g., religious associations, civic groups, race- and ethnically-based groups, consumer interest groups, environmental associations, gay and lesbian interest groups), as Iris Marion Young has argued, do not necessarily and primarily work for the "collective self-interest of their [own] members."[27] These associations often coalesce with one another; and the issues they advance often echo, or have bearings on, the concerns of other associations; and consequently, the issues they raise impact the society as a whole. In direct-deliberative *e*-democracy, associations of this sort could also function as important apparatuses of maintaining pluralism as well as securing social cohesion. The same is also true of the conception of "associative democracy," advanced by Cohen and Rogers, that focuses on economic associations (e.g., unions and professional associations) as the main vehicle of deliberative democracy.[28] In addition to participating at macro levels in the capacities speculated above, associations in direct-deliberative *e*-democracy can also utilize deliberative-democratic modes of decision-making at micro and meso levels, either within their own organizations or within the context of coalitions they form with similar types of organizations and associations. This could be done regardless of whether they are social or civic (in the case of Habermas' conception) or economic (in the case of Cohen and Rogers), in addition to participating at macro levels in the capacities stated above.[29]

Equality and the Question of Property Relations

As developed in Chapter 1, the "conception of democracy as the political empowerment of the citizen" is coupled with a conception of equality that is somewhat thicker than the one espoused by liberal democracy. Chapter 1 treated the question of equality primarily as an issue belonging to the realm of the political, and in a weaker sense, to that of the social. This treatment took equality primarily as *equal power sharing* or as the *equal political and social empowerment of all*. However, on the question of economic equality or distributive justice, the "conception of democracy as the political empowerment of the citizen" remained largely silent. Despite this shortcoming of the concept, the discussion in Chapter 1 tentatively assumed that a somewhat higher level of equality of the economic circumstances than the one existent in present-day American society would be necessary for the actualization of the idea of democracy as the political empowerment of the citizen. Throughout its long history up to the late nineteenth century, democratic theory was always coupled with a strong component of economic equality. This association was so strong that, in order to underscore it, Macpherson characterized democracy as a "class affair" and as a "leveling" and "equalizing" concept.[30]

Now, notwithstanding the fact that the notion of economic equality has historically appeared as indispensable to democratic theory, one can argue that the notion should not be regarded as a foundational element of the theory. Rather, it should be viewed as a co-condition or a companion of the concept, insofar as the question of the actualization of the idea is concerned. There are no logical necessities for coupling democracy with economic equality. Despite the fact that the theory of direct-deliberative *e*-democracy clings steadfastly to the original meaning of the idea of democracy as citizens' sovereignty (and conceptualizes it as the individual's political empowerment via individuating sovereignty), it distances itself from this companion that often tends to overshadow the true meaning of the idea of democracy—and consequently promotes an understanding that regards democracy primarily as a "leveler" or "equalizer" in the realm of material holdings.

Thus, going against the grain of "classical theory," the theory of direct-deliberative *e*-democracy takes democracy more as an equalizer in the social-political sphere and less as a "leveler" in the domain of material holdings. Despite this disclaimer, nonetheless, the theory assumes a state of affairs in which the existing economic inequalities are adequately ameliorated. For the notion of social-political equality to be substantive, this state of affairs would be necessary. As was noted earlier in this chapter, this severing of the egalitarian implication of democracy from the concept itself in the theory of direct-deliberative *e*-democracy is a consequence of developing the theory within the liberal-democratic framework, which takes democracy solely as a collective practice of social decision-making. Moreover, as will become apparent in the rest of this section, this de-coupling of democracy and economic equality would prove to be strategically indispensable to the goal of realizing the idea of direct-deliberative *e*-democracy.

This turning away from the egalitarian implications that have historically been associated with the concept of democracy is indeed a source of difficulty for the theory of direct-deliberative *e*-democracy and could turn out to be its theoretical Achilles' heel if it is not dealt with carefully. On this point, the theory of direct-deliberative *e*-democracy is vulnerable to the same set of criticisms that the liberal-democratic conception has been defending itself against throughout its short history. One should recall that liberal democracy became possible, both theoretically and in practice, by discarding the egalitarian implications of democracy, as well as by grafting onto itself a de-substantialized concept of sovereignty—(i.e., a version that is indirect, mediated, discontinuous, and intermittent). Despite their similarities in dealing with the question of economic equality, the main difference between liberal democracy and direct-deliberative *e*-democracy, as well as the advantage of the latter over the former, lies in that direct-deliberative *e*-democracy steadfastly holds on to the original notion of sovereignty and gives it a true substance by placing the power to exercise it directly at the fingertips of citizens. Moreover, by doing so, direct-deliberative *e*-democracy also gives substance to liberal democracy's formal conception of political equality—often expressed by the motto of "one person, one vote." Given that direct-deliberative *e*-democracy's notion of the individual's political empowerment assigns "equal weights" to all individuals' political wills (regardless of their social and economic standing), and also given that all individuals are empowered to exercise their sovereignty *directly*, its conception of political equality turns out to be essentially a principle about *exerting sovereignty equally*. In this respect, direct-deliberative *e*-democracy's conception of political equality is as substantive and as actual as it is theoretical and formal—(a claim that cannot be made about liberal democracy's conception of political equality in actual or practical terms).[31] In the case of direct-deliberative *e*-democracy, it is the individual citizens' direct participation in exerting sovereignty that guarantees political equality.

Beyond these general comments, one now needs to address the question of how the theory of direct-deliberative *e*-democracy should deal with the existing widespread economic inequalities, and how it could prevent these inequalities from becoming an obstacle on the path to its realization. The institutional arrangements proposed in Chapter 3 only depicted a stage at which the Realistic Democratic Utopia had become fully operational, without saying much about its initial conditions or the transitional period—which could take a few decades. There, the assumption was that, by the time direct-deliberative *e*-democracy would become fully operational, the question of distributive justice would have been adequately addressed and a realistic system of fair and equitable property relations would have been established. Given the significant role played by large-scale private property and wealth in shaping and directing political power in present-day America, it would be complete folly to assume that the Realistic Democratic Utopia could develop to its final stage without satisfactorily addressing the question of economic equality, let alone to assume that it could function as a stable system of governing in its fully operational mode.

Therefore, one should assume that during the transitional period, the progress toward the fully functional stage of direct-deliberative *e*-democracy should necessarily be accompanied by a parallel progress toward addressing the question of distributive justice adequately *and* achieving a more equitable society as well. As a matter of fact, one can reverse the question and argue that direct-deliberative *e*-democracy could be instrumental in pressing society to address the question of distributive justice, and to do so in a realistic and pragmatic fashion. That is to say, by setting in motion the process of transition toward the fully functional Realistic Democratic Utopia, one would also necessarily inaugurate a gradually progressing and comprehensive public dialogue on how to expand the scope of the citizens' participation in political decision-making—this dialogue would necessarily encompass the question of addressing the existing economic inequalities. At any rate, for reasons that will become apparent shortly, the public discourse on expanding democracy should either take place chronologically prior to, or serve as the medium for, the public discourse on expanding economic equality and addressing the question of distributive justice.[32]

* * * * *

But what shape or form should the parallel process of addressing the question of distributive justice assume? Admittedly, discussing the question of economic transition is a much harder task to accomplish than that of discussing the process of political transition. For one thing, the former process would follow the process of political transition and, thus, its parameters would be set by this transition, at least in its beginning stages. Another important reason for this difficulty is that there is no precedent for public dialogues on the question of economic equality in the prevailing public political culture of present-day society. The notion that the question of economic justice could ever be addressed in large and open public venues appears much more utopian to the political consciousness of this society than the idea of giving citizens more power in social decision-making than they presently have.

The mass media in the United States are replete with pseudo-discussions, articles, and commentaries on the need for more democracy (or the need to give private citizens a louder voice in government). These media are also saturated with a continuous theme of the need for decreasing the influence-peddling of big money and lobbyists in politics, especially during the election seasons. The ubiquitous talk of "campaign finance reform" is a case in point. It goes without saying that these pseudo-discussions in the mass media are mainly intended as channels for citizens to vent their anger and dissatisfaction with the political system, as well as being intended as campaign slogans for the first-time candidates running against the incumbents. However, no such pseudo-discussions or sloganeering ever surface in the popular media on the question of distributive justice or economic equality. For these reasons, it is difficult, and unnecessary at this point, to speculate how a manageable public debate on addressing the question of distributive justice could be introduced into the public political culture of present-day America.

Having granted this, however, it is reasonable to assume that, in this debate, at least in its beginning stages, the question of distributive justice will be meshed with that of democracy at the workplace. Thus, it is reasonable to assume that the need for addressing the question of distributive justice would arise most likely out of the public debates on furthering the citizens' democratic empowerment at the micro and meso levels such as the workplace.[33] It is also reasonable to assume that a substantive type of workplace democracy would require some form of incrementally increasing employee stock ownership as a way of bolstering employees' economic stakes at the workplace. These stakes would not only provide incentives for employees to participate, but would also gradually raise their social and political powers at the workplace. This program of economic and socio-political transition at the micro level could serve as a basis for instituting a macro model of economic transition that would proceed shoulder to shoulder with political transition toward the fully operational stage of direct-deliberative *e*-democracy. Higher degrees of direct democracy at macro levels would mirror higher degrees of economic equality and direct democracy at the workplace and vice versa.[34]

* * * * *

To bring this section to a close by way of summary, the theory of direct-deliberative *e*-democracy separates the concept of democracy from the idea of economic equality with which it has historically been coupled. On this, the theory of direct-deliberative *e*-democracy follows in the footsteps of liberal democracy. However, unlike the case of liberal democracy, for which this separation is motivated by positing democracy and economic equality as conceptually independent of each other, the separation in direct-deliberative *e*-democracy is motivated mainly by *strategic* considerations. Despite severing the concept of democracy from economic equality, the theory of direct-deliberative *e*-democracy takes the latter as being a *co*-condition of the former, as well as taking both as being strongly compatible with each other. Thus, in contrast to liberal democracy, direct-deliberative *e*-democracy requires a certain degree of economic equality as a *co*-condition for its full realization. The path leading to this state of affairs would be a gradual process of political transition that would require a complementary program of economic transition. This program would be necessary for the stability and smooth functioning of direct-deliberative *e*-democracy in its fully realized stage.[35]

However, setting the economic equality requirement as a *pre*-condition for direct-deliberative *e*-democracy would, no doubt, hinder the initiation of the process that would lead to its full realization. The same can be said about the idea of setting limits on unequal distribution of incomes as a way of reducing economic inequalities or any other radical measure that could make reaching a public consensus on implementing direct-deliberative *e*-democracy difficult. This is mainly due to the great influence the propertied classes in the United States presently exert over the government, media, and the political scene in general. Thus, in its initiating stages, direct-deliberative *e*-democracy should provide a full constitutional guarantee of private property rights, as well as some constitutional clauses to protect these rights against being restricted for a certain period of time (e.g., a decade or two), and that thereafter, the introduction of such restrictions should be gradual. A good starting place for reforming property-relations would be the existing inheritance rights and laws.[36] For the sake of launching the new order and maintaining its stability, until direct-deliberative *e*-democracy is fully operational, citizens of direct-deliberative *e*-democracy would be constitutionally forbidden from treating private property as a political issue. This also means that the citizens would not have the constitutional power to take up the issue of the socialization of large-scale private property.[37] Nevertheless, there is no question that once direct-deliberative *e*-democracy consolidates itself, the issues of economic equality, in all of their various forms and shapes, will inevitably find their way into the arena of public debate. By then, one should assume, the economic transition program has kept up with the political component and this debate will not prove to be de-stabilizing. Finally, as should be transparent by now, the primacy given to the discourse of democracy over that of economic equality here is primarily a question of *political strategy and pragmatism* and not that of the theoretical subjugation of the question of economic equality to that of democracy.[38]

The "Guardians of the Public Interests" and the Question of the Quality of Voting Outcomes

As has become evident by now, the faith of the theory of direct-deliberative *e*-democracy in the moral and intellectual competence of ordinary citizens to make sound political decisions is a qualified one. This is the main reason why the theory requires a solid component of representation. Recalling from an earlier part of this chapter, direct-deliberative *e*-democracy was characterized as a strong and vibrant *representative democracy* as well as a system of the *power-sharing and partnership of the elites and ordinary citizens.* The latter characterization is an indication that the elites (the guardians and representatives) are as essential to the theory of direct-deliberative *e*-democracy as the citizens themselves. Although citizens express their wills directly, and these expressions are entered into the decision-making process without any mediation, the process in which these wills are formed is not unmediated at all. Rather, as has been discussed on numerous occasions, this process is influenced greatly by the public discussion and deliberation sessions and other informative public forums directed by the guardians that precede each voting occasion.

Notwithstanding this intermediary educational role played by the guardians, one needs to reiterate that in direct-deliberative *e*-democracy, the ultimate power and the responsibility of will-formation rests with individual citizens themselves. Moreover, the actual decision-making power also rests with the individual citizens, both directly and indirectly. Indirectly, insomuch as they themselves elect the guardians (and the representatives), who not only influence the citizens' will-formation process, but also vote alongside the citizens; and that the citizens also have the power to unseat the guardians in the next election, if they choose to do so. Directly, in that citizens also directly partake in voting and contribute their share to the decision-making process. Furthermore, depending on the rate of their voting turnout, votes cast by citizens could potentially outweigh those of the guardians and the representatives combined. This means that citizens of direct-deliberative *e*-democracy are endowed with the direct power needed to nullify and reject political programs and policy options advocated by elites, and even revolt against them, by voting against their recommendations using their protest option and veto power.

The need for guardians in the theory of direct-deliberative *e*-democracy stems from the plausible and realistic assumption that, left to their own devices, not all citizens would have the high levels of moral and intellectual competence needed to enable them to contribute positively to the collective decision-making process, and also that not all of those who are competent would stay at these levels of moral understanding, reasonableness, or knowledge at all times. Democracy of the magnitude represented by the theory of direct-deliberative *e*-democracy cannot function properly without a highly knowledgeable and educated body-politic; and such a body-politic cannot exist without an extensive system of social-political and citizenship education that begins early in life and continues into adulthood. In the theory of direct-deliberative *e*-democracy, the guardians are conceptualized first and foremost as the "directors" and "facilitators" of the self-education of the adult public on the social and political issues that face the public. In order to make informed and prudent decisions, it is necessary that citizens be civic-minded, generally knowledgeable, and well-informed on each particular issue put before them for voting. In this respect, the idea of guardianship is an indispensable component of the theory of direct-deliberative *e*-democracy.[39]

In the theory of direct-deliberative *e*-democracy, the guardians are conceptualized as representing the citizens' aspirations and commitments to bringing their public wills to conformity with the moral and rational claims relevant to the issues put to them for voting. In other words, the guardians are conceptualized as representing both the social-moral conscience and the ideally high social-political consciousness of citizens. The guardians are also conceptualized as being well-versed in matters of social-economic policy, either by their superb education and research (e.g., as leading social-political thinkers, philosophers, economists, and social and political scientists), or by the exemplary service they have provided to their communities. Moreover, they are also judged by citizens as possessing these qualities, having strong moral characters, and being trustworthy as well. In the persons of the guardians, citizens would see the near perfect images of themselves as ideal citizens. The guardians are also conceptualized as representing citizens' aspirations for embedding social-political decision-making and political leadership in knowledge and integrity. The guardians are expected to bring knowledge, reason, morality, and virtue to politics, not just by partaking directly in decision-making in their respected assemblies, but also by directing the processes and institutions that facilitate the citizens' political self-education and inform their will-formation activities.[40] Moreover, they are also needed to act as intellectual and moral counterweights against the anti-liberal popular sentiments or anti-democratic tendencies, which could flare up among some individual citizens in times of crisis. The guardians would serve these functions both directly and indirectly: directly, in that they would engage citizens dialogically in deliberations or informal discussions; indirectly, insomuch as they would provide leadership for the institutions devoted to the goal of facilitating or enhancing the political and civic self-education of the citizenry. Guardians would also serve these functions by being sources of knowledge and social-political wisdom, to which citizens can turn and consult in the process of forming their public wills.

Despite what a *prima facie* understanding of the idea of guardianship might lead one to believe, the guardians in direct-deliberative *e*-democracy are not of the same species as the Platonic "Philosopher Kings." Dissimilarities between the two can be sketched in terms of the following considerations. First, unlike the guardian philosophers of Callipolis who subjugate ordinary people to their own interpretations of moral and rational principles, the guardians of direct-deliberative *e*-democracy only facilitate the self-education of citizens on these principles—they do so by guiding, motivating, and persuading individual citizens to bring their thoughts, wills, and political behaviors to conformity with these principles in all matters public (and perhaps private as well). Moreover, unlike the Platonic guardians who impose upon ordinary people the policies and decisions that are guided by these principles, the guardians in direct-deliberative *e*-democracy join citizens to make these decisions collaboratively. Finally, unlike the Platonic philosopher kings who indoctrinate or manipulate the citizens of Callipolis by propaganda and noble lies, the guardians of the Realistic Democratic Utopia attempt to keep citizens vigilant and politically active by cultivating in them the powers of analytical-critical thinking, and self-criticism as well.[41] Unlike Callipolis, which is the realm of "the sovereignty of the philosopher king," as Popper puts it, the Realistic Democratic Utopia is the true reign of the sovereignty of citizens.[42]

The dual functions assigned to the guardians so far (i.e., the function of decision-making in their assemblies, and the function of leading the deliberative institutions, and directing the institutions that facilitate the self-education of the citizenry on public issues) put them in the position of being intermediaries who link together the "civil society," on the one hand, and the "political society" (the state), on the other hand. These dual functions of guardians also place upon their shoulders the responsibility of advocating and protecting the "general interests" of the "civil society" against the interests of the "economic society," which would be championed by the "representatives."[43] Two important components of fulfilling this responsibility would be those of maintaining and nurturing a democratic political culture, and generating and fostering the "social capital" needed for the sustenance and nourishment of this culture.[44] This would be both an educational task, insofar as the guardians would need to "culturalize" the public and direct the self-education of the citizenry, and a political one in that they would need to take on the "representatives" in order to secure the financial, physical, and human capital needed for their educational and cultural work.[45]

The (independent-virtual) "representatives" in the theory of direct-deliberative *e*-democracy, on the other hand, are conceptualized as representing the interests of the "economic society" in particular, and as standing for this-worldly aspects of the citizens' needs and wants, in general.[46] The representatives are conceptualized as the men and women of this-worldly affairs and pragmatic knowledge and wisdom. As successful business people, attorneys, and professionals, their political perspectives on issues and policies, and thus their input into the decision-making process, would be guided by their commitment to the material well-being of their constituents and the society writ-large. Just as the guardians who would inform the public and persuade citizens to consider the issues and policies from the standpoints of moral considerations, knowledge, and the long-term interests of society, the representatives would do the same, but mainly from the standpoints of the practicality, and the short-term and material interests of the citizens and society. Moreover, given their backgrounds and their functions, the representatives would also play the intermediary role of linking together the "economic society," on the one hand, and the "political society" (the state), on the other hand. In other words, in many respects, "representatives" in the Realistic Democratic Utopia would function very much in the same way as they now do in present-day American liberal democracy. Using the U.S. system as an example, the representatives in the Realistic Democratic Utopia would function as the members of the House of Representatives and Senate currently do—with the exception that in the Realistic Democratic Utopia, there would be no pretense that these individuals (or most of them) have deep social-philosophical and social-scientific knowledge, or that they possess superb moral understanding, or have the non-market-related public service credentials that the guardians would be expected to have in the Realistic Democratic Utopia.[47]

As has become transparent by now, the contrasts drawn between the functions to be performed by the guardians and the representatives, on the one hand, and between their professional-educational backgrounds and qualifications, on the other hand, reflect the contrast and the dualism that permeates social life.[48] This dualism is nothing other than the problem of balancing public interests with private interests that modern liberal democracies and their conscientious citizens grapple with on a daily basis. In direct-deliberative *e*-democracy, this dualism will be balanced by both addressing the problem publicly and attempting to reach compromise solutions collaboratively. In order to assure that the guardians will perform their function in this crucial balancing of interests successfully, they will be required to be free of party and market affiliations. This requirement is intended to guard the guardians against the influence-peddling of the economic society. This requirement is also premised on the contention that political parties, broadly speaking, are formed around special economic interests and their political programs and social philosophies necessarily echo these interests.[49]

Given their functions, the representatives, in contrast to the guardians, will have party affiliations. As was speculated in Chapter 3, the full institutionalization of the concept of guardianship and the inauguration of the assemblies of guardians would most likely be among the final measures on the path toward fully actualizing the idea of direct-deliberative *e*-democracy. In the transitional period, the guardianship functions of leading the deliberative institutions and directing the self-education of the public would be largely performed by well-known and respected leaders of the civic associations. These activities of the civic leaders, who would be in essence the forerunners of the future guardians, would take place solely within the civil society (without any involvement in the legislative process) and would be publicly funded.

* * * * *

Now, having emphasized that direct-deliberative *e*-democracy is both a representative democracy and a system of power sharing of citizens and the social-political elites, it is worth reiterating that the main function of the guardians is to enhance the quality of the citizens' contributions to the collective decision-making process. This function would be performed primarily through their roles as the directors of the institutions and processes that facilitate or guide the political self-education of citizens on issues put to the public vote. In addition to serving in this capacity, the guardians would also affect the voting outcomes in other ways. In order to prevent the over-politicization of society and for reasons of efficiency and practicality, not many issues would be classified as "major" issues.[50] This would mean that most of the issues in direct-deliberative *e*-democracy would be decided mainly by the guardians and representatives. Moreover, the guardians and the representatives would also function as the agenda-setters for the debates and issues to be put to the public vote. The guardians would bring to light the moral and technical-factual aspects of the issues to be decided, whether "minor" or "major." These functions of the guardians would greatly enhance and influence the quality of the decisions reached. In the case of the "minor" issues, the quality of outcomes, on the one hand, would directly reflect the quality of the assemblies and the values affirmed by the constitution of the direct-deliberative *e*-democracy itself; and, on the other hand, would be an indirect reflection of the values affirmed by the citizens themselves who have elected the members of the assemblies. In this respect, direct-deliberative *e*-democracy is on par with liberal democracy.

In the case of "major issues," on the other hand, the quality of outcomes, i.e., the quality of decisions made through composing the "collective wills," would be a direct reflection of both the values of citizens and the assemblies, as well as a reflection of the institutional values of direct-deliberative *e*-democracy itself. Viewed in this light, a decision reached by the process of composing "collective wills" on a major issue ought *not* to be regarded as the aggregation of the (self-interested) preferences of citizens (though their self-interests have been accounted for in the process). Rather, such a decision should be viewed as a reasonable and morally justifiable decision and, at the same time, as the judgement of the whole society rendered collectively and fairly on the issue in question.[51] The justification for this claim lies in that the process of composing the collective will out of the inputted wills of citizens is but the final stage of a longer process that begins with agenda-setting and preparing policy options, and then proceeds to the public deliberation stage, wherein citizens subject their opinions and preferences to the scrutiny of factual knowledge, moral considerations and reasonableness, and develop their political wills. In this long process, the institutional values of the society are brought to bear on the agendas, policy options, and wills of citizens. Thus, when a citizen of direct-deliberative *e*-democracy rejects or endorses a policy option (whether she does so fully or partially) in deciding on a major issue, it is not the case that she simply expresses her self-interested preferences; rather, she judges whether the policy option in question corresponds to the institutional values of society and, at the same time, passes judgement on whether that policy option is in line with the interest of the public. In light of these considerations, the voting outcomes reflect the values affirmed in the constitution *and* the "public political culture" of direct-deliberative *e*-democratic society. (The question of the "public political culture" in the direct-deliberative *e*-democracy will be addressed in the following two sections.)

Justifying Direct Democracy

Now, the conceptualization of direct-deliberative *e*-democracy as a system of power-sharing of the citizens and political elites, needless to say, runs counter to the theory and practice of the modern constitutional representative democracies. The history of these democracies so far has been, in one sense, the history of the rule of the political elites over the rest of the society. This rule has always been justified in the name of the people. The liberal-democratic conception has contributed much to the actualization of this rule as well as furnishing moral and rational justifications for it. Moreover, as was argued in *The Betrayal of an Ideal*, the idea of representative government in the modern world has often been offered as a "wise" alternative to the "unwise" idea of the direct participation of ordinary people in governing. *The Betrayal of an Ideal* categorized all of the assumptions that have historically been used as premises for arguing against the idea of direct democracy into seven presuppositions and offered a short examination of some of them. These presuppositions were referred to as the *assumptions of the impracticality, inefficiency, and undesirability* of the idea of ordinary citizens participating directly in governing, and the *assumption of uninterestedness of ordinary citizens* in doing so.[52]

As an integral part of developing the theory of direct-deliberative *e*-democracy, one needs to counter these presuppositions that justify the rule by political elites and, at the same time, argue that the idea of the citizens' direct participation in political decision-making is not as "unwise" as it has been made to appear. This will be done in what follows immediately. *The Betrayal of an Ideal* categorized these presuppositions as follows:

I. *The logistic impracticality and inefficiency of direct rule by the people*: the immense size and complexities of the modern nation-state make it impractical (and inefficient), if not impossible, to assemble the entire citizenry for policy- or decision-making purposes.[53]

II. *The political impracticality or inefficiency of direct rule by the people*: the facts of the heterogeneity of the citizen body and the diversity of perspectives, and the existence of a wide plurality of conflicting interests in the modern nation-state, make practicing the idea of direct democracy inefficient, if not impossible.

III. *The political-technical and intellectual incompetence of ordinary citizens*: ordinary citizens lack the political-technical knowledge and the intellectual-rational power needed for sound and consistent social-political decision-making. This problem is compounded in the face of the ever-rising complexity of governing in the modern nation-state.[54]

IV. *The moral incompetence of ordinary citizens*: Ordinary citizens lack the moral competence needed for sound social-political decision-making.[55]

V. *The inability of ordinary citizens to discern their own true interests and the true interests of the whole community*: ordinary citizens lack the wisdom, fortitude, and other necessary attributes needed to raise them above the level of their immediate interests and concerns and to help them develop a true understanding of their common interests.[56]

VI. *The political uninterestedness of ordinary citizens*: ordinary citizens are politically apathetic and lethargic. They are not interested in participating in politics and prefer to leave the responsibilities of governing to professional politicians.[57]

VII. *The susceptibility of ordinary citizens to fall for demagogues and the fear of instability*: a system of direct-participatory democracy would either be an unstable republic on the permanent verge of collapse, thanks to constantly being undermined by the ordinary citizens' whims and prejudices, or would eventually fall prey to demagoguery.[58]

Beginning with presupposition VI (*assumption of the uninterestedness of ordinary citizens*), and limiting the scope of the discussion to the American scene, there is no denying that this presupposition is based on empirical observations. However, one can argue that this presupposition, notwithstanding the fact that it is based on empirical evidence, misrepresents the facts about the prevailing political apathy or the lethargy of the citizens in this society. Far from being evidence of the political uninterestedness of citizens, one can make a good case that these observations ought to be interpreted as a manifestation of the citizens' political alienation from a political system that has been consciously designed to keep them away from the political process.[59] "They [citizens] are apathetic because they are powerless," as Barber observes.[60] Thus, these empirical observations should be regarded not as a testimony to the political apathy or uninterestedness of the citizens, but rather as a testimony to the success the representative democracy has exhibited in its attempts to keep the citizens away from the business of governing. Empirical evidence aside, one can argue that presupposition VI finds theoretical support in the overall attitude of liberalism toward the individual and its postulation of citizens as mutually disinterested.[61]

In attempting to weaken presupposition VI, one can turn to empirical evidence of a different sort. The rise in popularity and the use of ballot initiatives, especially in the southwestern states in the United States in recent decades, is one such piece of evidence. Another piece of evidence is the rise of the idea of "citizen activism" or "activist citizen" that gained currency in the 1960s and early 1970s. In recent decades, this idea has contributed to the mushrooming of numerous movements (the so-called new social movements) and organizations that actively seek greater involvement in politics and advocate the individual citizens' political participation in politics. The major movements and organizations in this spectrum are devoted to issues of the environment, gender, race, ethnicity, sexual orientation, consumer interests, and profession-based interests (e.g., unions of white-collar workers). Finally, one can also weaken presupposition VI further by pointing to what David Miller calls "the decline in authoritarian values" in late-twentieth century America.[62] The sentiment that we should "throw the rascals [the professional politicians] out" surfaces among the public very often. In sum, presupposition VI no longer carries the force it did in a not-so-distant past.

Now focusing on presuppositions III-V (*undesirability assumptions*), one can argue that these presuppositions, no doubt, are based on degrading appraisals of the private citizens' capacities and potentials. Presuppositions III and V are, in essence, pseudo-epistemological assumptions about ordinary citizens. Presupposition IV, on the other hand, is a pseudo-moral justification for keeping ordinary citizens out of politics. In arguing against these presuppositions, one can take a three-pronged approach.

On the first prong of this argument, one can contend that these assumptions are metaphysical in nature and that their appeal to present-day intellectuals is largely due to the immense influence of the negative theories of human nature and the degrading attitudes toward everyday people that have loomed large in the intellectual tradition of Western civilization.[63] Moreover, one can further argue that these presuppositions, as well as the theories of human nature that bestow intellectual respect on them, have aristocratic roots and class-based biases—indeed a quality that has overwhelmingly tainted much of the intellectual production in the realm of social-political thought throughout the history of Western civilization. Given this state of affairs, and on behalf of direct-deliberative *e*-democracy, one would need to argue for a new metaphysics, a paradigm shift, and a theoretical reorientation on the question of the private citizens' political and moral capabilities. This can be argued for on the strength of the positive theories of human capacities, such as those developed by Macpherson and Marx, and also on the basis of the fact that the level of the general public's education has risen considerably during the latter part of the twentieth century. Admittedly, achieving this metaphysical shift will not be an easy task given that one must go against the intellectual-cultural grain of Western political thought that has consistently derided and scorned the intelligence and moral capabilities and characters of private citizens.

Putting the metaphysical aspects of the *undesirability assumptions* aside, on the second prong of contending against these presuppositions, one can argue that political decision-making does not require high levels of technical expertise or extraordinary reasoning powers. Here, one can borrow the argument Barber has advanced against the "defenders of the rule of expertise." Those who doubt the ability of the public to make judicious decisions, in Barber's view, betray their misunderstanding of how the legislatures function. According to Barber,

> the legislative function . . . is not to institutionalize science or truth but to judge the public effects of what passes for science or truth. Citizens are not different from elected legislators in this regard: their task is to judge, evaluate, and assess—*to employ judgement rather than expertise*. . . . Political judgement above all involves evaluating options in terms of value priorities, and as such it is available to every woman and man willing to submit their personal opinion and private interests to the test of public debate and political deliberation.[64]

Lastly, on the third prong of this contention, one can argue in favor of the feasibility and utility of moral-intellectual and citizenship education of the citizens as a way of weakening presuppositions III-V. The guardianship component of the theory of direct-deliberative *e*-democracy has been intended primarily for countering these presuppositions. Moreover, the guardianship component has also been intended to incorporate some aspects of the Platonic idea that knowledge, morality, and virtue ought to be the bedrock of political decision-making. Finally, the guardianship component has also been intended to give expression to the Rousseauean idealization of democracy as a system of education of the citizenry.

Now, taking presuppositions III-VI together, one can further argue that these presuppositions are more sociological and cultural, and perhaps psychological, in nature than philosophical. And consequently, they ought to be countered by relying on the results produced by the vast research conducted in these fields to show that moral and citizenship education do alter the political attitudes and behaviors of citizens. As can be recalled from Chapter 3, direct-deliberative *e*-democracy takes a cautious and pragmatic approach to the idea of directly involving citizens in political decision-making. The strategy there was to gear the intensity, degree, and tempo of the process of the gradual involvement of citizens in direct decision-making to empirical research that would study the various aspects of experimenting with citizens' direct involvement at various levels of decision-making.

Finally, returning to an earlier theme in this chapter (that the faith of direct-deliberative *e*-democracy in ordinary citizens and their moral-intellectual competence was a qualified one), one should add that this faith is directly related to the interconnections between practicing political participation by citizens, on the one hand, and their moral competence and knowledge, on the other hand. On the interconnections between knowledge and political participation, Barber puts the matter quite eloquently when he states:

> Give people some significant power and they will quickly appreciate the need for knowledge, but foist knowledge on them without giving them responsibility and they will display only indifference.[65]

Now focusing on presupposition II (the second of the *impracticality and inefficiency assumptions*), one can argue that this presupposition is in essence a theoretical-ideological assumption that feeds primarily on suspicions and fears that the ever-growing political pluralism and cultural diversity of the contemporary American society could undermine its unity and political stability. Furthermore, one can argue that this presupposition overstates the case for the problems inherent in managing the diversity of perspectives and addressing the conflict of interests. As to the difficulty of managing public deliberations in a culturally pluralistic (multi-cultural) social milieu, one can argue that, as a way of weakening the appeal of presupposition II, it is possible to channel a wide range of perspectives into a manageable manifold of views around some platforms that would be agreeable to most and tolerable to others, provided that an appropriate (i.e., accessible, equal, and fair) public forum for deliberation and addressing the questions of social stability and civic solidarity could be created. Of course, this could be a difficult task to face at times. It could well be the case that platforms for deliberation and discussion could not be formed as often as one would hope; and even if they were formed, it is possible that participants would talk past each other rather than to one another. Nevertheless, keeping afloat an ongoing dialogue (whether direct or indirect) among diverse and competing conceptions of the "good life" and "good society" (common good), as well as giving ample opportunities of expression to those who do not choose to enter into any form of dialogue with others, is essential to the stability of a culturally pluralistic society.

At the minimum, such public forums would help achieve two things: expose the participants to the "moral and political perspectives" of other cultures, hence enriching their understanding and appreciation of the values in other cultures; and give under-recognized or under-represented views or cultures a fair hearing, and thus prevent them from becoming marginal or peripheral.[66] Moreover, public forums devoted to such goals could also help to bring out the limitations of the political system and hint at the ways the system could be improved.[67] Finally, these forums, one can argue, would help instill the desired quality of "reasonableness" in some of the extreme comprehensive religious, moral, or political doctrines that find their way into American society. Canada's successful experiment with cultural pluralism (multiculturalism) in recent decades is a testimony that such stabilizing general platforms do exist.[68] In direct-deliberative *e*-democracy, the endeavors for assuring this aspect of social stability and civic solidarity will be directed by the guardians.

But how about the question of political stability and the question of collective will-formation in an environment of *both* political and cultural pluralism? Recalling from Chapter 3, the direct-deliberative *e*-democracy would be replete with a wide spectrum of parties and associations that would promote diverse and competing conceptions of the "good life" and "good society." Along with promoting these conceptions, the parties and associations in question would also promote a wide range of diverse and competing political platforms that spring from these conceptions and are deemed conducive to their realization. While political parties, broadly speaking, would seek political power mainly around economic platforms and interests, the civic associations, on the other hand, would do the same on the basis of their commitments to matters and issues that are of concern mainly in civil society. Among these, one can mention the recognition and empowerment issues related to ethnicity, religion, race, gender (e.g., NAACP and NOW in the United States), and non-economic special interest issues (e.g., the case of the National Rifle Association in the United States). One should keep in mind that civic associations do not necessarily put their cases as special interests in every instance. In some cases they represent issues that are of interest to all, such as environmental and consumer protection issues, or in the case of societies that promote physical well-being (e.g., the American Cancer Society) or cultural and educational issues.[69] While lying somewhere between the parties and the civic associations, workers' unions and professional associations (e.g., those of teachers and engineers) would also compete for political power. Given this diverse political environment, the political heaven of the Realistic Democratic Utopia would be lit with constellation after constellation of diverse political opinions and views. Thus, one should assume that this situation would result in the formation of numerous "opinion blocs"—(loosely formed around the diverse and competing conceptions of the common good and public policies on particular issues)—that would be advanced by various parties and associations. These opinion-blocs would also form around "special interests." Given this currency of numerous and diverse political views, it would be reasonable to assume that none of these opinion blocs would have a dominant or majoritarian character or status.[70]

Now, the question to be discussed is as follows: in such a diverse political environment, how can one expect to reach policy or legislative decisions or form collective wills on major issues when citizens are involved directly? The answer to the question lies in the specific collective will-formation method discussed in Chapter 2 (FEFI). Given that this method offers a diverse and meaningful set of policy options to choose from, and given that it allows voters to enter their votes in practically infinitely many different ways, one can argue that the opinion blocs would begin to dissolve as soon as the voters start to enter their votes. There is no good reason to believe that the opinion blocs in question would translate directly into "voting blocs," unless one assumes that most voters would vote "irresponsibly" or "ideologically." And this would not be a reasonable assumption.[71] The case for the assumption of widespread "irresponsible" voting behaviors grows even weaker if one further makes the reasonable assumption that no responsible political party, or no civic or professional association, especially the ones with higher social status and wider membership, and thus with higher stakes, would encourage their supporters or members to vote irresponsibly. The best the political parties and associations could do would be to try to convince their supporters to consider the views expressed by them seriously, and hope that they would get a large portion of their supporters' 300 points on the policy option(s) they support. Given this scenario, the winning policy option (i.e., the collective will formed) would not be a collation of some opinion or voting blocs, but a composition of wills expressed by individual voters who had considered these opinions and had voted on the basis of their best judgements, expressed in their own unique ways. If no policy option were to win clearly, which would seem to be the most likely scenario, then the decision would have to be made in a runoff vote by channeling the wills into one of the two policy options with the strongest showing in the first round.

What is of utmost importance in this politically and culturally diverse social-political environment is that the stability and viability of the "public political culture" be assured, and a strong sense of unity and common-collective identity be forged. These indeed would be formidable tasks that would be left to the guardians. This question will be revisited later in this section.

Now the case of presupposition I (the first of the *impracticality and inefficiency assumptions*). Of all the presuppositions discussed above, presupposition I is the only truly objective one (and a real point of concern for the idea of direct democracy) in that it does not misrepresent the facts and that it is free of ideological prejudices and theoretical biases. Having granted this, one should be quick to point out that presupposition I is premised on a certain historically conditioned and limited understanding of the practice of direct democracy that can be stated as follows: decision-making in a direct democracy takes place in physical (i.e., spatial-temporal) assemblies wherein the decision-makers (citizens) assemble together and hold face-to-face deliberations that eventually lead to decisions, via the use of some sort of voting scheme (or by seeking consensi). This is in fact the model of Athenian direct democracy that was also practiced in some isolated pockets in medieval Europe.[72] There are a number of problems with this understanding of the practice of direct democracy in contemporary society. To begin with, the immense size of the modern nation-state makes the ancient Athenian model of practicing direct democracy impractical. Moreover, the idea of holding face-to-face discussions amongst citizens that would culminate in a collectively made decision on the spot is an "ideal" that cannot be actualized in large venues such as the national or state levels in modern nation-states. Nor is this ideal practicable in large-to-medium-sized urban settings. As a matter of fact, on the basis of the original meaning of the idea of democracy, one can argue that the idea of face-to-face discussions is not an indispensable component of direct democracy. Rather it is only one form of practicing it. The main merit of the face-to-face ideal is that it brings the whole community together, or at least a good part of it, and thus helps strengthen the communal bonds.

Notwithstanding this merit, this "ideal" form of direct democracy faces a major difficulty, i.e., the community gatherings could turn into public forums for compulsive speakers and would-be demagogues, wherein they would rob the attendees of their valuable time. It is this negative side of the "ideal" form of Athenian direct democracy that moved Rousseau to claim that "Athens was in fact not a Democracy, but a very tyrannical Aristocracy, governed by philosophers and orators."[73] The other merit of face-to-face interactions is its interactive feature that facilitates the exchange of ideas at interpersonal-dialogical levels. However, one should add that this feature of face-to-face assemblies should be regarded in a positive light only if one assumes that all the participants are extroverted. In fact, one can argue that decision-making in face-to-face assemblies would be oppressive for introverted individuals present in the crowd, who would not feel comfortable in voicing their views in public. Face-to-face deliberations favor those who are extroverted and, in addition, are skilled in formal debates and have the ability to argue their points dispassionately.[74]

Returning to the contention that face-to-face deliberations in physical assemblies ought not to be taken as a pre-condition for practicing direct democracy, and that such assemblies are not practical at macro levels, and perhaps not desirable given that they could become "tyrannical," presupposition I begins to lose its force of appeal and cogency. Furthermore, one can argue that face-to-face assemblies would be very inefficient and could put a high demand on the citizens' time by taking "too many evenings" and thus could have discouraging effects on mass participation.[75] Moreover, one can argue that the large-scale assemblies could have disempowering and alienating effects on the introverted individuals, who find themselves overwhelmed and psychologically strained in large crowds. Now, having brought to light these shortcomings of face-to-face interactions, one can argue that the latest communications technologies could remedy some of them. Electronic townhall meetings or video conferencing in direct-deliberative *e*-democracy would function as virtual assemblies. These assemblies could be nationwide or restricted to specific localities. Participation in these assemblies would be interactive, and the exchange of ideas would take place by the participants calling in or sending electronic messages to the assembly that would then be read to other participants.

Admittedly, the community-bonding effects of these virtual assemblies would be minimal. Nevertheless, these assemblies would be as interactive as the actual assemblies, with the extra advantage that they would also enable the introverted individuals to express their views. Moreover, there would be no danger of having the assembly hijacked by the compulsive orators.[76] Finally, another advantage of using the latest *e*-technologies is that they make it possible to hold round-the-clock and lengthy online debates and discussions, whether in real-time, such as chatrooms, or in virtual-time (i.e., by exchanging electronic mails or posting ideas on electronic bulletin boards and Internet sites). To sum up the discussion of presupposition I, this presupposition points to a major difficulty in realizing the idea of direct democracy in contemporary large nation-states. However, as was argued, the latest innovations in communication technologies can overcome these difficulties.

Lastly, the case of presupposition VII. In arguing against this presupposition, one should point out that this presupposition is in effect a conclusion that is drawn on the strengths of presuppositions III-VI, and in this sense is the most metaphysical of them all. Hence, in arguing against this presupposition, it suffices to contend that, in light of the arguments presented earlier against presuppositions III-VI, presupposition VII would also lose its force and cogency. A second way to confront this presupposition is to view it within the context of the Realistic Democratic Utopia of Chapter 3, and reiterate the arguments presented there, to the effect that direct-deliberative *e*-democracy would have built-in features to prevent demagoguery.[77] Lastly, speaking within the context of the theory of direct-deliberative *e*-democracy again, one should be reminded that direct-deliberative *e*-democracy is not a pure system of direct democracy, but one that has a vibrant representative component and a system of checks and balances that would protect it against potential demagogues.

Having weakened the appeal of these presuppositions, especially that of presupposition I, one can now argue for the cogency of direct-deliberative *e*-democracy on the grounds that the democratic utilization of the latest innovations in the technologies of communication and the mass media have made a new form of direct democracy possible. The macro assemblies of citizens in televised electronic townhall meetings directed by the guardians, as well as the individualized debates and discussions that can take place in the virtual plane among citizens, or between citizens on the one hand, and their representatives and guardians, on the other hand, could be as effective as the Athenian town assemblies insofar as the question of discussing the issues and developing understandings about them is concerned. Participation in these assemblies, whether direct or indirect, would help the individuals to "discover the truth" in the Rousseauean and Habermasian sense. As was discussed above, one disadvantage of electronic assemblies in comparison to the Athenian-style face-to-face town meetings was its poor utility in helping to bond the community.

This shortcoming of electronic assemblies, as will be seen in the following section, would be compensated by some of the functions played by the guardians. Despite this shortcoming, one can argue that the direct-deliberative *e*-democratic type of assemblies has the advantage of being able to avoid some of the pitfalls of the Athenian model, such as inefficiency, demands on the time of the citizens, and the susceptibility to being hijacked by compulsive talkers and would-be demagogues. Moreover, on the question of efficiency in casting the ballots and the ability to fully express one's wills, direct-deliberative *e*-democracy appears far superior to the Athenian model. Lastly, for the reasons cited above and other obvious reasons, practicing democracy directly at macro levels in the style of Athens is now a practical impossibility. Consequently, the only feasible avenue open to practicing the idea of direct democracy in contemporary society is the direct-deliberative *e*-democratic way.

The Public Political Culture and Civil Society

Returning to the earlier contention that in direct-deliberative *e*-democracy, voting outcomes would reflect the values affirmed in the constitution and the "public political culture" of the society, one might ask the following questions: what are the interrelations between the institutional values of direct-deliberative *e*-democracy and its public political culture? Where do the guardians fit in this picture? Moreover, do the values that guide citizens' thinking on public issues—(and condition their will-formation process)—differ from the ones upheld by the guardians and representatives, or from the ones affirmed in the constitution of direct-deliberative *e*-democracy? To answer these questions, one needs to begin with the contention that what holds direct-deliberative *e*-democracy together in this realm is its "public political culture." This culture is sustained and promoted, as well as developed and nurtured, by the guardians, who do so in conformity with the underlying principles of the constitution of direct-deliberative *e*-democracy.

The task of promoting and developing the public political culture of direct-deliberative *e*-democracy, as well as keeping citizens vigilant and well-versed in it, is the third function of the institution of guardianship. While the guardians would attempt to direct citizens toward viewing the issues from the relevant moral and rational standpoints (the broader good of the community being the focal point), the representatives, on the other hand, would address citizens' concerns regarding their material well-being and self-interested pursuits with their brand of this-worldly knowledge and prudence. Moreover, as was discussed earlier in this chapter, the dualism underlying the contrasting functions of the guardians and the representatives in this respect is a reflection of the dualism of the social life that manifests itself to the individual in the opposition between the public interests and her private ones. The three-layered voting scheme of Chapter 2 was designed with one eye on addressing this deep-seated dualism that permeates the social life of the individual. In public deliberations, the guardians would contribute to enriching and deepening the citizens' understanding of the policy issues on the first two layers, whereas the representatives would do the same for the third layer.

The "public political culture" in direct-deliberative *e*-democracy would be part and parcel of its "ideology" and value system. In light of the image portrayed of direct-deliberative *e*-democracy so far, one way to characterize this society would be to state that it is to be understood ideally as a vast and pluralistic community (or a nation) of citizens, who have united around a set of elastic and publicly accepted conceptions of the "good life" and the "good society" (as well as around a shared (thin and yet broad) conception of fundamental universal values and ends regarded as worthy, a common ground, an "overlapping consensus" indeed), and have committed themselves to honoring a set of fundamental universal principles that sustain and nurture these conceptions.[78] These fundamental principles are but the ones affirmed in the constitution of direct-deliberative *e*-democracy. The citizens' allegiance to these principles, and their adherence to the ideology and culture that promote them, as well as their deeply felt sense of civic and moral responsibilities to honor these principles, constitute their "common or collective identity" amidst the differences that divide them.[79] As the sustainer of society's publicly accepted conceptions of the "good life" and "good society," as well as being their legal manifestation, the constitution of direct-deliberative *e*-democracy is committed to the well-being of the individual, as well as to the well-being of the community itself. The former commitment is materialized via granting the individual a rich package of basic rights and liberties, both positive and negative. The latter commitment, on the other hand, is actualized through the institution of guardianship that would inspire, assist, and impel citizens to honor their civic responsibilities and commitments to the community wherefrom they draw their rights, liberties, and powers.[80] One should also add that the institution of guardianship also plays a significant role in fulfilling the commitment of the society to the well-being of the individual. It does so by providing moral and intellectual guidance to the individual in drawing up her life plans, and in using her rich package of rights and liberties effectively in order to fulfill these plans.

This discussion brings up a familiar controversy between liberalism and communitarianism (or civic-republicanism for that matter) over whether the interests of the individual or the community should take priority over the other. One can argue that, within the context of what was just argued, this controversy loses its significance and becomes irrelevant in direct-deliberative *e*-democracy. While the constitution of direct-deliberative *e*-democracy grants the individual a rich package of rights and liberties, and places the sovereignty and political power at his fingertips; at the same time, it gives the guardians ample educative powers to keep the collective interests of the community present in the consciousness of the individual. Moreover, the constitution, and along with it the public political culture and its highly-esteemed guardians, advises the individual to assign a heavier weight to the interests of the community than to his personal interests when he exercises the political power at his fingertips.[81] Thus the question, "Which takes priority over the other, the individual or the community?" cannot be answered in direct-deliberative *e*-democracy, for the relation between the individual and the community is a fluid one. As a result, the question of priority within this context becomes indeterminable in principle.[82] This fluidity is directly linked to the conception of the individual presented in Chapter 1, which postulated the individual as a fluid unity of self-regarding concerns and social-orientedness, and wove social consciousness into the fabric of the individual's identity.

Now, returning to the importance of the notion of "public political culture" in direct-deliberative *e*-democracy, one needs to add that this notion underscores the importance of the concept of "civil society" to the theory of direct-deliberative *e*-democracy. The civil society in direct-deliberative *e*-democracy can best be conceptualized as the totality of those societal spheres, institutions, entities, and activities wherein, and through which, the public political culture is maintained, nurtured, and developed. As defined, it then follows that the concept of civil society being employed here deviates somewhat from its received view that takes the concept, broadly speaking, as the sum total of all societal entities and spheres that are apart and distinct from the state (the political society). In contrast to the received view, civil society in the theory of direct-deliberative *e*-democracy is taken to mean the totality of all those societal structures and spheres that are not only apart from the state but are also distinct and apart from the "economic society."

In other words, here, civil society is taken to encompass all those societal structures, domains, and activities that are distinct from both the state and the "market" and wherein, or through which, citizens come together on a regular or customary basis. These structures, domains, activities, and the institutions that support or enhance them are of great importance to the public. Some examples of civil society here would be schools, churches and similar religious institutions, universities, libraries, museums and institutions for the promotion of arts and culture, and the "local talk shops."[83] The civil society in this formulation also includes those institutions and activities that are developed and planned specifically for the purpose of assembling the citizens for discussing local community issues and concerns, or for socializing around these issues. Relevant examples here would be townhall meetings, citizens' boards and action groups, neighborhood committees and meetings, "civic homes," and civic associations.[84] Some segments of the mass media of communication would also fit here, but only to the extent that they are publicly owned or publicly operated, or are genuinely devoted to public interests. To these one should also add the social space of the workplace, provided that there, citizen-employees have the time or opportunity to mingle, socialize, and exchange ideas on social-political issues.

Broadly speaking, the major portion of the institutions and domains of civil society are open and accessible to all citizens.[85] Moreover, the operations and activities of these institutions and domains are guided by a set of norms or principles that are egalitarian in nature, at least ideally. (Egalitarianism here, in part, results from the fact that the public's participation in most institutions of civil society is voluntary.) In comparison with the institutions of the "economic society" and the state, the structures of most institutions of civil society are either non-hierarchical or semi-hierarchical. Furthermore, compared with the state and the economic society, especially the latter, (a well-functioning) civil society nurtures—and at the same time is nurtured by—the values of "trust," "reciprocity," "sympathy," "mutual support," "fellowship," "good will," "cooperation," and "social networking," which "social capital" theorists have been keen on emphasizing in recent years.[86] Moreover, the mode of communication in the civil society is either interactive, e.g., meeting and socializing in person (or doing the same in real time on the Internet); or indirect, e.g., through the mass media of communication. Finally, in most institutions and domains of civil society, broadly speaking, the public has considerable power and leverage to intervene directly and affect changes, e.g., local schools.[87]

Now, what is excluded from the civil society *and* the state, and lumped together as the "economic society," are those entities, elements, and spheres of society and the workplace that are outside the domains of the public's direct influence or its direct exercise of control. Simply put, the economic society is the domain of society's economic activities in a "free-market" economy. Among these, one should include the economic firms and corporations, their administrative and ownership structures, their boardrooms, as well as the corporate culture and its "ideology" and world outlook (which nurture and promote the core values of competition, productivity, efficiency, and profit). Moreover, one should also include the relations of production and distribution, capital itself as a social-economic force, and the privately owned media that operate as for-profit organizations.

Viewed from the standpoint of the individual, what separates the civil society from the economic society is that the economic society is the domain where he *works* (primarily for making a living) and the civil society is the societal sphere where he *lives*, *enjoys* his life, and *develops* his human potentials.[88] This latter characterization reveals that civil society also encompasses some aspects of the domain of the private life of the individual. Finally, one should mention the state (the political society) as the third and last sphere of society. The state is the sphere in which the civil society meets the economic society. The state, thus, is the intermediary sphere where the civil society's and economic society's claims against one another are mediated—and eventually reconciled in the form of laws.

Now, bearing in mind this division of society into the trio of civil society, economic society, and political society (the state), one can argue that, in present-day America, the economic society occupies the center stage. The economic society in the United States dominates and conditions both the civil society and the state. As to the relations between the political society and the economic society, it is common knowledge that the financial power of the economic society greatly conditions both the political processes and their outcomes, including legislative ones, to the extent that the state takes on the role of the stewardship of the economic society.[89] Apropos of the relations between the economic society and civil society, one can argue that not only the former regiments the workplace, and structures the life activities of citizen-employees in accordance with its own needs and ends, but it also invades and dominates the free space of the civil society.

Speaking theoretically and in broad terms, the civil society belongs to no one and to everyone at the same time. However, in practice in present-day America, the civil society lives under the sway of the economic society's "culture" and "ideology" (i.e., its conceptions of the "good life," "good society," and the "public good"). This domination of the civil society by the economic society shapes the world outlooks and the moral-cultural norms of the citizens in ways that serve the interests of the economic society. Moreover, this dominion also conditions the moral-political character of intellectual and artistic productions in favor of the economic society, and thus prevents a free play of intellectual and cultural developments in the civil society. All successful attempts to escape this domination, be they political-intellectual movements, artistic works, or entertainment, are eventually sabotaged, subverted, or bought off, and brought under the control or patronage of the financial power of the economic society. A good way to characterize the power relations among the trio would be to resort to Gramsci's concept of *hegemony*.[90] That is to say, to assert that the culture of the economic society is hegemonic in present-day America.[91]

In contrast to present-day society, the civil society in direct-deliberative *e*-democracy would move gradually to the center stage. The distinctions between the state, civil society, and the economic society would become somewhat blurred in the fully realized stage of direct-deliberative *e*-democracy. This would be mainly a direct consequence of both the citizens' political empowerment in the workplace and their participation in the legislative functions of the state. This blurring of the distinctions among the trio would also be in part a consequence of instituting the conception of guardianship and redefining the functions of the representatives. Given their functions, the guardians would have one foot in civil society as "educators" and civic leaders, and another in the state as legislators. Meanwhile, the representatives would perform their own dual functions with one foot in the economic society and the other in the state.

The Question of Technology

The theory of direct-deliberative *e*-democracy relies on a specific category of the *e*-technologies that have mushroomed in the course of the last two decades or so. The theory identifies three potentials in these technologies and utilizes them for its democratic purposes. First, the theory of direct-deliberative *e*-democracy uses communication-media technologies as effective tools for conducting or facilitating public deliberations and for promoting civic education.[92] Second, it utilizes the latest computer and information-communication technologies toward the goal of making voting a logistically simple and inexpensive operation to undertake, thus making a case for more frequent use of voting as a mechanism for collective decision-making.[93] Finally, the theory of direct-deliberative *e*-democracy employs electronic voting technologies in order to expand the utility of voting beyond its current use of merely registering the monosyllabic yes/no inputs of voters.[94]

Undoubtedly, the theory of direct-deliberative *e*-democracy would not have been possible without these technologies. The very formulation of the theory is not only premised on the existence of these technologies but is also inspired by them.

The reversal of the old adage "where there is a will, there is a way" best captures the relation between the theory of direct-deliberative *e*-democracy and the latest *e*-technologies: "where there is a way, there is a will." This relation can also be explained by an intuitively appealing and empirically tenable, as well as an historically documented, and consequently a self-evident, general thesis about the impact of technology on society that can be expressed as follows. New technologies and new technological capacities give rise to new conceptualizations of what is to be done in the social contexts in which they arise, as well as giving rise to new ways of doing the old things. New technologies give rise to new ends to be set for realization. They also alter or transform, or become instrumental in transforming, their social contexts.[95]

This general thesis adequately explains the relation between the idea of direct-deliberative *e*-democracy and the latest *e*-technologies in question. This relation, one can argue, does not obligate the theory of direct-deliberative *e*-democracy to espouse a particular philosophical system on the larger question of technology. Nor does it require the theory to develop its own supplementary theory of technology. As was argued in Chapter 3, direct-deliberative *e*-democracy does not require its own specific set of technologies. Nor does it call for the further technologization of society. Rather, it takes the existing *e*-technologies and their established structures as they are, and puts them to a new use. Moreover, the theory of direct-deliberative *e*-democracy does not thrive on the claim that the latest *e*-technologies have altered the overall social and political well-being of the American society for the better. Surely, these technologies have contributed much to improving economic productivity, brought about new opportunities for economic development, and revolutionized the uses and capabilities of the communication and information networks. However, they have also invited serious criticisms and have raised concerns that call into question the wisdom of embracing them in blithely unconcerned and uncritical manners. These technologies have also raised questions apropos of how, or even whether it is possible, to foresee the long-term moral, social, and political effects of the transformations they are bringing about, almost on a daily basis.

These questions have been raised and discussed for over a decade as they have joined some larger questions on technology that began to surface in the middle parts of the twentieth century. The questions then stemmed from concerns regarding the adverse and uncontrollable effects of what appeared at the time as the total technologization of society. These questions constitute the subject matter of the thriving field of the philosophy of technology and its related sub-fields. The philosophy of technology is primarily guided by inquiries into how technologies transform social attitudes, and how they both deconstruct and reshape our conceptions of self, values, and even the reality itself. The philosophy of technology is also guided by questions that seek to study the ways in which technologies alter social-political institutions, affect the nature of political power, and shift its balance in the society.[96] A discussion of these questions falls out of the realm of the problem at hand and thus will not be considered here. Instead, it will simply be contended that since these technologies are widely in use, and have already become part and parcel of everyday life in the United States, they could be utilized for democratic purposes in a radically different way than they have been used thus far. The theory of direct-deliberative *e*-democracy starts with the fact of the omnipresence of the new computer and communication technologies, their ease of use and accessibility, and the fact of their integration into the everyday life activities in the United States, and proceeds to offer a theoretical framework for their utilization in the service of furthering the cause of democracy in this society. In light of these considerations, and also given that it only relies on a specific category of technologies and utilizes them in a limited capacity, the theory of direct-deliberative *e*-democracy is not obliged to address the "question of technology" writ-large.[97]

Notwithstanding this disclaimer, however, given that the theory of direct-deliberative *e*-democracy utilizes *e*-technologies for political purposes, it has to address a set of questions which put forth the arguments that the utilization of technologies can, and does, produce (or become instrumental in producing) anti-democratic political structures and adverse social-political consequences.[98] Among these, one should mention the contentions that technologies facilitate the administration of citizens by governments; that they "acculturate [citizens] to compliance and conformity," strip them of the ability to make "principled-decisions," and undermine their "will to reason"; that they replace real communities with "pseudocommunities" and "pseudorealities," and as a consequence, create a citizen body that has diminished capacities for grasping the public context and social nature of the everyday life, and appreciating the need for "reciprocity" and human interactions; and finally, that technologies glorify expertise and scientific knowledge, which in turn, downgrade and devalue the citizens' personal knowledge.[99]

The counter argument here would be that, far from aggravating the problems these arguments bring to light, the system of governing proposed by the theory of direct-deliberative *e*-democracy can be conceptualized as a technology of governing that would empower citizens and help reverse the adverse political impacts of the existing modes of utilizing technologies in present-day American liberal democracy.[100] The decision-making process in direct-deliberative *e*-democracy, one can argue, would draw, in due proportions, on both the knowledge and prudence of the experts *and* the personal knowledge of the citizens. Moreover, in direct-deliberative *e*-democracy, the culture of conformity of the present technological society would give way to the culture of participation through technology. Rather than being controlled or becoming the objects of the decisions and plans made by the experts, as is the case now, citizens in direct-deliberative *e*-democracy themselves would become the controllers, the planners, and the makers of principled decisions. Finally, one can argue that direct-deliberative *e*-democracy would ameliorate most of the problems associated with the phenomena of "pseudocommunities" and "pseudorealites," albeit only to the extent that it is feasible in a vast and complex society. The sense of political power that each citizen would feel, and the sense of social responsibility that would accompany this power, in concert with a public political culture centered on civil society, would help spawn lively public venues at all levels, and in all corners, of society. These venues would provide ample incentives and reasons for individuals to step out of their pseudo communities and embrace real ones, and thus achieve true communion with others in the actual world of shared living.

* * * * *

Having briefly considered the political question of the relations between technology and democracy, the theory of direct-deliberative *e*-democracy now needs to address a series of specific concerns and potential criticisms that have been brought forth in recent years (or could be voiced in the future) on the question of making democracy, or the practice of politics in general, dependent upon the latest *e*-technologies. As will be seen below, these criticisms and concerns are *technical* in nature and directly target the electronic voting aspect of the theory of direct-deliberative *e*-democracy. Given that these concerns and criticisms are often expressed in direct and concrete terms, they will need to be addressed in direct and specific terms as well. These criticisms can be categorized into the following five propositions:

I. *The problem of securing electronic voting against sabotage and infiltration.* It is possible for unscrupulous individuals or groups, or underground dissident organizations, or foreign governments, to sabotage the system, or affect the outcomes of voting by infiltrating it. A good analogy here would be the case of computer hackers breaking into the Internet stores and stealing the credit card numbers of customers.

II. *The problems of fraud in voting.* It is possible that some individuals can access other individuals' voting PINs and vote in their names.

III. *The problem of protecting the privacy of the voters and the confidentiality of the votes.* The voters' political inclinations and voting patterns may lose their confidential or private characters (in the same way that the consumption patterns and behaviors of consumers have become public information and put in the service of targeted advertisement). This might discourage some from voting online.

IV. *The problem of the "digital divide," also known as the problem of the "access divide" and the "inequitable access" to information and communication technologies.* The electronic voting scheme would discriminate against the less well-off or the poor, e.g., poor blacks in the United States, who cannot afford to own computers.[101]

V. *The problem of digital illiteracy and computer "skill divide"*[102]. The electronic voting scheme would discriminate against those who are computer illiterate or have limited computer skills, mainly the poor and the elderly.[103]

As one way of addressing the concern expressed in criticism I, it can be argued that votes cast on the Internet can be compiled by local electronic servers or computers and checked for irregularity, such as viruses or evidence of tampering, before being forwarded to a central compiling computer. If this solution is unsatisfactory, or the scheme of voting via the Internet cannot be technologically secured against sabotage, the other possible solution would be to use a publicly charted and operated closed net system, instead. The closed net system could be accessed via electronic voting machines that can be placed in secured rooms in public places such as libraries, schools, colleges, and community centers that would be staffed and operated during the voting time/day blocks (approximately once every three months). Electronic voting machines are already in existence. And despite concerns and questions regarding their security and accuracy in the months leading up to the November 2004 elections, they were used extensively and, in most cases, successfully, on Election Day. The question no longer seems to be whether electronic voting machines will acquire universal usage and become standard voting tools, but how they can be made reasonably secure and more accurate.[104]

In order to assure accessibility during the voting day/time blocks, the portable versions of electronic voting machines could be taken to those locations where the residents are unable to travel to voting sites, e.g., nursing homes and community centers.[105] Even without the Internet, the scheme of closed net electronic voting would be far superior to the existing system in terms of the accessibility of voting booths, ease of voting, and the ease of tallying the results. Once the voting time block expires, the data in these machines would be transferred electronically to the "democracy computer" for tallying the votes and tabulating them. In order to assure that the software in these machines has not been tampered with and the voting outcomes would not be skewed, these machines and the data stored in them would be checked for irregularities (such as viruses) before the data was transferred to the main computer. Finally, as an extra measure of security, electronic voting machines could be programmed to print out two cards right after the voter pushes the appropriate button to cast or send her vote. These cards would be similar to the ballot cards dropped in the ballot boxes in the existing system, with the exception that they would clearly indicate in print how the voter has voted, in addition to being punched or bar-coded appropriately. The voter then would drop one of these cards in the ballot box and keep the other for her own records. The main reason for this measure is to leave a "paper trail" in case the voting results are in doubt or disputed. The paper trail can also be used to check the accuracy of the electronically obtained results at individual voting districts, at random.[106]

In responding to criticism II, one can maintain that this is a non-issue. Even given this possibility, this pattern of voting is more secure than the existing practice that does not require any form of I.D. at the voting booth. Moreover, the security measure of a PIN is presently being used for ATMs and credit cards with a satisfactory security record. Lastly, one can argue that in the near future, the PIN system will be replaced by devices that would establish the identity of the individuals at the ATMs by using their fingerprints, voices, or the patterns of the pupils of their eyes. The same sort of security can be used by the electronic voting machines. As a response to criticism III, one can argue that the second solution suggested for criticism I above could also counter this criticism. (Information on voters' choices recorded by the public closed net system would be destroyed after the voting results are certified.) In the case of criticism IV, one can argue that this problem is becoming less of an issue with the passage of time.[107] This is mainly due to the fact that computers are becoming more affordable. Moreover, one can argue that the second solution suggested for criticism I would also be applicable here. The third argument here would be to say that the direct-deliberative *e*-democratic society's commitment to democracy and social equality would compel it to take aggressive measures to solve the problem of the "digital divide" early on in the process. Finally, one could argue that criticism V should not be regarded as a serious contention against direct-deliberative *e*-democracy. The problem of computer illiteracy or skill-deficiency is similar to the problem of literacy in its usual sense, which was a major issue when universal suffrage was instituted in the United States last century. The fact that a good portion of the population was illiterate was not a serious argument against instituting universal suffrage and expanding democracy. The remedy then was to begin a campaign of literacy and assist the illiterate in the voting process. The same sort of solutions would also work in the case of expanding *e*-democracy in general, and in using electronic voting machines, in particular.[108]

Concluding Remarks

This chapter brought together the "conception of democracy as the political empowerment of the citizen" (developed in Chapter 1), the theoretical framework of the "Realistic Democratic Utopia" (explored in Chapter 3), and the scheme of amalgamating-composing collective wills (developed in Chapter 2) and synthesized them into a single theory that was referred to as the "theory of direct-deliberative *e*-democracy." The synthesis began in the course of comparing the new idea with the liberal-democratic conception, on the one hand, and the deliberative-participatory theories of democracy, on the other hand. The theory of direct-deliberative *e*-democracy was then fully elaborated within the context of discussing numerous key questions, including those of democratic legitimacy, the citizens' direct participation in politics, economic equality, and technology.

From liberal democracy, the theory of direct-deliberative *e*-democracy appropriated its constitutionalism (proceduralism, commitment to the individual's rights and liberties, and its notion of formal political-legal equalities). From deliberative democracy, on the other hand, the theory acquired its appreciation for the epistemic and morally educative values associated with the idea of public deliberations, and thus its deliberative component. Finally, and most importantly, from participatory democracy, the theory of direct-deliberative *e*-democracy appropriated the idea that, far more than being just a "free method" for electing "legitimate" governments, *democracy is essentially a moral concept.*

In the Conclusion, it will be argued that the theory of direct-deliberative *e*-democracy is capable of working toward resolving or dissolving some of the important questions that have dogged political philosophy for the longest time.

Notes

1. The choice of the name for the theory will become apparent throughout the chapter. For the time being, it suffices to say that direct-deliberative *e*-democracy is "direct" in that citizens directly participate in decision-making; it is "deliberative" in the sense that decision-making is preceded by public deliberations; finally, it is "*e-democratic*," in that the theory utilizes electronic media.

2. As has been argued in *The Betrayal of an Ideal*, Macpherson has shown that the ideal of democracy has historically been coupled with strong egalitarian ideals. To emphasize this, Macpherson has characterized democracy as a "class affair." Taking the case of Rousseau's conception as an example, Rousseau required a strong principle of material equality: no person should be rich enough to buy another person; and no person should be so poor as to be bought by someone else.

3. Two notes are in order here. First, the phrase "public political culture," and the general idea behind it are borrowed from Rawls (Rawls 1993, p.175). Second, two important features of the framework in question are constitutionalism (proceduralism, an unswerving commitment to the individual's rights and liberties, and a strong notion of formal political-legal equalities), and the belief in the separation of the private and public domains. Among other features of this framework, one should include "tolerance," "reasonableness," "common sympathies," and "moral nature" and other attributes of the "liberal people" (See Rawls 1999, pp.23-24, p.29, and pp.47-48).

4. As was argued in Chapter 1, the package of rights and liberties was taken as a set of pre-given political and historical facts, as well as representing one of the highest political achievements of human civilization, enjoying the approval of the overwhelming majority (if not all) of the individuals in present-day American society—thus reflecting its tradition (or its "facticity" as Habermas would put it, e.g., Habermas (1996), p.xxvi and p.454-55), as well as being acceptable to the participants in the Rawlsian original position.

5. See Principle (XII) and a related discussion in Chapter 3.

6. This proposition is similar to Dahl's "Presumption of Personal Autonomy" (Dahl 1989, p.99). Ian Budge argues that the presumption of personal autonomy is the "utilitarian argument" for democracy (Budge 1996, p.9). According to Budge, this argument has been "well exemplified in contemporary economics"—"that the individual's expressed preferences are the only legitimate indicator of his or her interest." However, for the reasons discussed in Chapters 1 and 2, the assumption 2 above should not be taken as being the same as the notion of "consumer sovereignty."

7. This assumption is similar to Nino's hypothesis that the "lack of impartiality is often due, not to selfish inclinations of the actors in the social and political process, but to

sheer ignorance of where the interests of others lie" (Nino 1996, p.119). Moreover, here, as was the case in the preceding chapters, by citizens, it is meant knowledgeable *political decision-makers* and not the self-interested rational-choice makers. Finally, this assumption is not as naïve as it appears. Surely, there are disagreements that arise directly from the interests of the groups and individuals. Material interests in the circumstances of sharp economic inequalities is the most important source of disagreement. As will be discussed later in this chapter, the full operation of direct-deliberative *e*-democracy is contingent upon somewhat ameliorated circumstances of inequality.

8. To be precise, the gap being bridged here is the one existing between negative liberties, on the one hand, and the first and third sense of "positive liberties" as defined by Macpherson, on the other hand. Macpherson breaks up Berlin's notion of positive liberties into three different concepts of (1) "individual self-direction or . . . self-mastery," (2) "coercion," and (3) participation, or as he puts it, "liberty as a share in the controlling of authority" (see Macpherson 1973, pp.108-09). While keeping the first and the third, Macpherson rejects the second. As a matter of fact, the whole "conception of democracy as the political empowerment of the citizen" developed in Chapter 1 corresponds to the third sense in full, and to some aspects of the first sense of positive liberties in Macpherson's typology of positive liberties. This contention will be further elaborated in the following pages in this section.

9. Some may argue that the public's participation in liberal democracy is also positive, for after all, it is the citizens who elect the officials into the office in the first place. This argument is not compelling at all. Given the fact that the representation in liberal democracy is virtual rather than actual, the public does not get to affect the decision-making process directly. Rather it only hopes that the elected officials will "do the right thing." It is for this reason that Riker characterizes liberal democracy, and rightly so, as a "minimal" democracy and sees the utility of voting only in this negative sense (Riker 1982, pp.9-10, p.246, p.253, and also pp.1-13).

10. Macpherson (1973), pp.108-09, also p.119. As can be recalled, it was argued in Chapter 1 that participation is in itself an act of self-development (Macpherson's first concept), and at the same time, it is an act that gives the participant a share in controlling her social milieu and transforming it as she sees fit (Macpherson's third concept).

11. Ibid., p.109.

12. Ibid. See Cunningham (2002), pp.36-39 for a discussion of Berlin's conception of positive liberties. Cunningham argues that espousing the first concept of positive liberty does not obligate one to also espouse the second concept: "to insist that people have the ability to select and revise the goals they follow and that this ability should be nurtured and protected [implied by the first concept] does not commit one to any conception of a higher and a lower self, much less to a theory about essential harmony among autonomously chosen goals [as implied by the second concept]" (ibid., p.37).

13. The following passage in Nadia Urbinati's "Democracy and Populism" is instructive: "[a]ccording to Tocqueville, from its inception, democracy in Europe emerged not as a *natural* fact, but as a *reaction against* the aristocracy. Much like Plato describes in *Gorgias*, equality in Europe has been used as an ideology of the many (the weak) to pull everybody down to a common level" (Urbinati 1998, p.113, original italics).

14. The following is Werner J. Dannuhauser's characterization of Nietzsche's attitude toward democracy in *The Encyclopaedia of Democracy* (Lipset 1995, vol.3, p.884):

> At rare intervals Nietzsche comments benignly on democracy, mostly because it counterbalances the corruption of authoritarian regimes.

> More often, however, his attitudes range from contemptuous indifference to vehement hostility. He frequently holds democracy responsible for the mediocrity that softens modern life and the egalitarianism that endangers human greatness. He faults it for being little but a prelude to communism. . . . Nietzsche tends to equate democratic rule with the rule of the last men.

15. The arguments presented against the representative government in *The Betrayal of an Ideal* are not tantamount to the outright rejection of the idea of representative government, but only to the rejection of the specific form in which it is utilized in the liberal and liberal-democratic states. The representative governments employed by these two states has two main democratic shortcomings: it is elitist by design and keeps ordinary citizens away from the policy and decision-making process, and it is susceptible to the influence-peddling of the "economic society" in general, and to the social strata that control or command it, in particular. *The Betrayal of an Ideal* characterizes these two shortcomings as "audience democracy" and "fund democracy," respectively.

16. A reference to Cohen (1997a), pp.84-86.

17. Here, common good is taken as the criterion of reason.

18. See Chapter 13 of *The Betrayal of an Ideal* for a discussion of the theories of deliberative democracy.

19. Habermas' version is a case in point. See Chapter 13 of *The Betrayal of an Ideal* for a discussion.

20. This orientation of direct-deliberative *e*-democracy, as was seen in Chapter 1, stems from its postulation of the individual as a fluid unity of self-orientedness and sociality. Again, as was seen in Chapter 1, the democratic conception that underlies the theory of direct-deliberative *e*-democracy was conceived not as an instrumental notion in the service of "rational will-formation" procedures or other similar goals with limited or concrete scopes, but rather as a total conception of the individual and her empowerment. True, the direct-deliberative *e*-democratic conceptualization of democracy can also be characterized as instrumental, but only in relation to the realization of the ultimate value of the human universe.

21. Nino (1996), p.106. Nino's version of deliberative democracy is discussed in Chapter 13 of *The Betrayal of an Ideal*.

22. Given that the citizen is both a social and socially conscious being, she is aware of the relations between her personal interests and the interests of the community, and thus is conscious of the necessity of considering the good of the community alongside her private interests. Therefore, when she expresses her wills directly, she does so at both the private and public levels. One should recall that the wills expressed directly by her have already been morally and intellectually "tempered" in the process of public deliberation.

23. As can be recalled, the last collective decision-making scheme proposed for macro levels in Chapter 2 can be conceptualized as a scheme of indirect negotiation among citizens in order to arrive at a compromise will.

24. Wills were discussed in detail in Chapter 1. It is worth reiterating at this point that the wills in the theory of direct-deliberative *e*-democracy are regarded as morally and intellectually tempered and reflected-upon judgements of citizens. These wills are expressed in the three layers of personal-interest, particular-public-interest, and general-public-interest at the moment of composing the collective-will.

25. The notion "expressive liberty" is Cohen's own (Cohen 1997b, p.419). The phrase "deliberative empowerment" is employed here to characterize what deliberative democracy intends to achieve.

26. See Chapter 13 of *The Betrayal of an Ideal* for a discussion of Habermas' version of deliberative democracy.

27. Iris Marion Young (1995), p.209.

28. See Cohen and Rogers (1993).

29. The "free associations" implicit in Habermas (discussed in his "Popular Sovereignty as Procedure")—(and one can assume that the same could be said about Cohen and Rogers' secondary associations, and Fishkin's deliberative polling scheme as well)—can be used to supplement direct-deliberative *e*-democracy.

30. See Part I (especially Chapter 2) of *The Betrayal of an Ideal* for a discussion of democracy as a "class affair." One should add that the thinness of the theory of direct-deliberative *e*-democracy on the question of economic equality becomes apparent only if one compares it to the "classical theory."

31. Here the allusion is to the second shortcoming of the liberal-democratic form of representative government. Despite the principle of "one person, one vote," the political wills of individuals with higher social and economic status have more weight than those of others in the decision-making process. By having resources to influence the process of nomination and the election of the representatives, and by having easier access to them once they are elected, the individuals with higher social and economic status exert more weight in the process than other citizens.

32. The notion that the discourse of the expanding democracy should be chronologically prior to the discourse of economic equality, in one respect, echoes Bowles and Gintis' linguistic conception of "postliberal democracy" in their *Democracy and Capitalism* (1986). According to Bowles and Gintis, "[w]hat is needed is the displacement of property rights by democratic personal rights" (ibid., p.178). A main building block of their conception of postliberal democracy is the notion of the "clash of rights," viz., the clash of personal rights vs. property rights, or the clash of the "expansionary logic of personal rights" and the "expansionary logic of capitalist production" (ibid., p.29). In their view, liberal democracy is incapable of offering a framework for the discourse of democratic personal rights that has heightened in late capitalism. Furthermore, they argue that Marxism, on the other hand, falls short of grasping the importance of personal rights, for it is "preoccupied with domination through unequal property ownership" (ibid., p.13); and its "theoretical lexicon does not include such terms as freedom, personal rights, liberty, choice, or even democracy" (ibid., p.19). Both "[l]iberal and Marxian social theories have conspired to depict personal rights as the indelible expression of bourgeois individualism" (ibid., p.152). Thus, the task of postliberal democracy should be to elevate the discourse of democratic personal rights to the position of dominance over property rights (ibid., pp.176-85).

33. As was stated in the concluding section of Chapter 1, the institution of the workplace democracy in effect would be the actualization of the micro principle of "social autonomy."

34. In foreseeing an economic transition program that could bring about a substantive type of workplace democracy, one can consider a thinner version of John Roemer's idea of "coupon or voucher socialism," expounded in his *A Future for Socialism* (Roemer 1994, pp.49-51, also pp.60-74). Briefly stated, beginning within the milieu of the existing system of property relations, this program could be started by mandating all large- and medium-size economic firms to endow their employees with some shares of the firm in

addition to their regular compensations. (One way to do this is to mandate that all or portions of the employees' pay raises be paid in the firms' stock shares.) This mandate could be supported and further encouraged by means of giving these firms tax-credits. These stock shares would entitle the employees to claim shares in the profits of the firm. The stock shares in question cannot be cashed in, but parts of them could be exchanged with shares of other firms. In its beginning stage, the stock endowments could be minuscule—or even symbolic in medium or smaller firms—but their value should increase gradually to a considerable amount in the course of a decade or two. Moreover, the firms would continue to be controlled and operated by managers. In a decade or so, as the process of political transition toward direct-deliberative *e*-democracy would approach the stage of full-actualization, the employee shareholders would become the main and dominant shareholders in firms. This growth in employees' shares would be accompanied by a proportional growth in their share of power in the firms, to the point that their appointees would weigh heavily in the boardrooms. Alongside this process, the employees' direct participation in administrating and managing the affairs of the firm at the micro and meso levels would grow in proportion. (The employees' direct participation in the micro and meso levels in effect would amount to the actualization of the principle of "social autonomy.") This program of economic transition would progressively decrease economic inequalities throughout the society. Moreover, the program would increase the employees' economic stakes in their firms, and along with it, their political stakes in the well-being and stability of society writ-large. It should be noted that Roemer's own proposal is not a transitional program at all. Rather it is a description of a fully implemented and elaborate model that amounts to a complete system of macroeconomics. The version of "coupon or voucher socialism" discussed here is thinner than Roemer's suggested program, which requires the massive involvement of government. In his original model, "the government distributes a fixed number of coupons or vouchers to all adult citizens" when they reach the age of 21 (ibid., p.49). And "[e]veryone's coupon portfolio would be returned to the public treasury at death, and allocations of coupons would continually be made to the new generation of adults" (ibid., p.50). Moreover, Roemer is of the view that this model could best be adopted in the post-communist countries of Eastern Europe, where the governments were already involved heavily in the economy (ibid., pp.126-27). Finally, the main difference between the thin version discussed here and Roemer's own version of this model is that his is designed in the plane of economics and lacks a democratic component, insofar as the question of the employees' indirect participation in the macro-management of the firms and their direct involvement at micro and meso levels is concerned. Cohen and Rogers (1996) and Jacobs (1999) have criticized Roemer's model for failing to extend democracy to the economic sphere.

35. Recalling from Chapter 1, Rawls' principles of justice were regarded as adequate guides for achieving the desired levels of economic (and social) equalities in the direct-deliberative *e*-democracy.

36. One should mention that Rawls' theory of justice also relies on taxing inheritance as a way of achieving economic justice (Rawls 1991, p.277f). Moreover, one should also add that in the *Communist Manifesto*, Marx and Engels' proposed program for radically altering property relations also emphasized reforming the inheritance laws, rather than advocating the public seizure of large-scale private property.

37. Carlos Santiago Nino's discussion of the interrelations between private property and democracy is most illuminating and relevant in this respect (Nino 1996, pp.157-60). According to Nino, "we must be careful not to produce social reactions that may undermine the whole democratic process" (ibid., p.160). This caveat concludes his concise

discussion of the question of private property and the history of its treatment in the American Constitution and jurisprudence, and its links to the British system of common law. Nino quotes Jennifer Nedelsky to the effect that, in this tradition, the issue of private property is regarded as "fundamentally legal, [and] not political" (ibid., p.158). The idea that direct-deliberative *e*-democracy ought not to treat private property as a political issue is inspired by Nino's discussion in these pages.

38. Barber advances a similar argument (albeit a weaker one) regarding the relations between democracy, on the one hand, and economic equalities, on the other hand. According to Barber, democracy must be first secured in its "thin and minimalist version," e.g., the model of the American democracy, and then be "reinforced as a strong democracy" by pursuing "social justice" (Barber 1998, p.39). Barber makes these comments in his "An Epitaph for Marxism," where he argues that Marxism failed because it was not practical; it was too idealistic, "too good for this world" (ibid., p.33), and "too removed from possibilities" (ibid., p.36). Alternately stated, Barber "places politics before economics" and argues that "only through civic revitalization can we hope, eventually, for greater economic democracy" (Barber 1984, p.305). The main difference between Barber's position and the one put forth by the theory of direct-deliberative *e*-democracy is that, while the former clearly places the question of social justice or economic democracy on the backburner, the latter puts the question of economic democracy in a parallel course with that of the political empowerment of citizens (or political democracy)—notwithstanding the fact that it assigns to political empowerment the leading role, which is mainly motivated by considerations of political strategy.

39. The view that democracy or a "democratically organized public" cannot exist without "social knowledge," "freedom of social inquiry," the "positive freedom" of "thought and its communication," and the "free and systematic communication" of the "conclusions" of social inquiry, is best argued by Dewey (Dewey 1954[1927], pp.166-84, especially pp.166-68, also p.208). For Dewey, "knowledge is communication as well as understanding" and that "knowledge of social phenomena is peculiarly dependent upon dissemination" (ibid., p.176). In the theory of direct-deliberative *e*-democracy, guardians function as the systematic disseminators and communicators of social knowledge. They perform this function both directly (by guiding discussions and debates) and indirectly (by leading the institutions that facilitate, organize, and disseminate social knowledge).

40. As was the case in the earlier chapters, each of the terms "reason," "morality," and "virtue" in this context is used in a broad sense. "Reason" is used to denote expert knowledge (i.e., technical and social-scientific (and to a lesser extent natural-scientific) knowledge, know-how, and modes of reasoning) relevant to social-political decision-making. Moreover, the term "reason" is also used to denote theoretical knowledge that understands the deeper social-philosophical issues and takes a long-term view of society and its aims and interests. The term "morality," on the other hand, stands for the importance of normative considerations or moral thinking in political decision-making, vis-à-vis the sort of influence the considerations of economic utility, efficiency, and monetary interests (often favoring special-monetary interests) can exert in social decision-making. Finally, "virtue" here underscores the importance of the need for truth, integrity, and sincerity in politics and political office. Furthermore, the notion of expert guardians in one sense falls along the lines of what Hilary Putnam terms "epistemological justification of democracy" in John Dewey's philosophy (Putnam 1992, p.180). Expert guardians can be regarded as, to use Putnam's phraseology, the "precondition for the full application of intelligence to the solution of social problems" (ibid.).

41. One can explain this latter difference between the two groups of guardians further by using the comparison Karl Popper draws between Socrates' "intellectualism," on the one hand, and Plato's, on the other hand. According to Popper, "Socrates' intellectualism was fundamentally equalitarian and individualistic," with only a "minimum" level of "authoritarianism," whereas Plato's was "the embodiment of an unmitigated authoritarianism" (Popper 1962, p.131). Moreover, Socrates' intellectualism was intended to "educate the citizens to self-criticism" (ibid., p.130). Furthermore, as an extension of this difference, one can argue that the *raison d'être* of the guardians in the direct-deliberative *e*-democracy is to enhance the quality of collectively made decisions, and not to "mediate between the divinity and society" by making the community, as Plato believed, "in the image of the divine exemplar," which the Platonic guardians were intended to do—phrases in quotation marks are borrowed from Wolin (Wolin 1960, p.54). Another obvious difference between these two groups of guardians is that the knowledge possessed by the guardians of direct-deliberative *e*-democracy does not have the absolute, static, unitary, and transcendental character (as in mathematics) which Plato seems to assume "true" knowledge would have. In this respect, the Platonic guardians were completely disconnected with, and independent from, the community they were to lead, and in this sense, the decisions they would make were not political at all. "By insisting that the political order must be linked to a transcendental order," as Wolin argues, "Plato could afford to ignore some of the most pressing issues of truly political nature" (ibid., p.51).

42. Popper (1962), vol.1, p.155.

43. A complete discussion regarding "civil society" will take place later in this chapter in a section titled "The Public Political Culture and Civil Society." For the time being, a brief outline provided in the second part of note 7 in Chapter 1 should suffice.

44. As with Chapter 3, the sense of "social capital" used here is the one defined by Putnam (Putnam 2000, pp.18-19). See note 51 in Chapter 3 for a brief statement on Putnam's usage of the term "social capital."

45. The role of the guardians in developing the democratic political culture will be discussed in the following two sections in this chapter.

46. Economic interests, broadly speaking, are particular or special interests, and in most cases, have local-regional or limited scopes, as well as direct or immediate impacts on the lives of the citizens at local levels.

47. The representatives' links to the market society would make them vulnerable to the influence-peddling of the market forces and their lobbyists, as is the case in the existing representative system.

48. In one respect the contrast drawn between the guardians and the representatives above reflects the contrast drawn by Plato (in *The Republic*, *Politicus*, *Laws,* and *Gorgias*) between "statesmen" or "philosophers," on the one hand, and the "politicians" and the sophists, on the other hand. While the statesmen aim at improving the moral-intellectual character of the citizenry, the "politicians" address the issues that relate to the citizens' wants, desires, and opinions with their own special skills (manipulation, deception, compromise-making, faction-building, etc.). (See Wolin 1960, pp.32-46, especially pp.45-46, for a discussion of this contrast in Plato.) The difference between the "representatives" in the direct-deliberative *e*-democracy and the "politicians" has to do with the reasonable assumption that, due to the influential presence of the guardians, "representatives" would use their political skills "honestly" and responsibly (e.g., striking fair and reasonable balances among competing interests).

49. As one can recall, some aspects of the functions of political parties in direct-deliberative *e*-democracy were discussed in Chapter 3.

50. Recalling from Chapter 3, there will be a formal division between "major" and "minor" issues. Citizens would also be able to contribute to decision-making in minor issues. Moreover, they would have the power to reclassify minor issues as major ones.

51. The question of why such a decision should be regarded as being made fairly was addressed in Chapter 2.

52. See Part II of *The Betrayal of an Ideal*, especially Chapter 10.

53. *The Betrayal of an Ideal* argues that his was the main thrust of J. S. Mill's argument against direct democracy (Mill 1861, p.55).

54. McLean (1989) discusses a different version of this presupposition which questions whether ordinary citizens have the political wisdom to avoid making "inconsistent" and "unwieldy" decisions, pp.109-11. The questions he poses challenge the prudence of a *pure* system of direct democracy. His solution to the problem is that the questions to be put to the public for voting must be presented to them as "complicated questions," p.111.

55. Here Schumpeter (1942) is a good case in point, e.g., p.257 and p.260. According to Schumpeter, people can easily be corrupted when it comes to pecuniary matters.

56. Again, Schumpeter is a good example here. The ordinary people lack a "command of facts," a reliable "method of inference," and "the sense of reality [in them] is so completely lost" (ibid., p.261). As was seen above, J. S. Mill had similar opinions of ordinary people.

57. *The Betrayal of an Ideal* argues that Schumpeter was a strong proponent of presuppositions III-VI.

58. For example., according to Schumpeter, people are "terribly easy to work up into a psychological crowd [even when they are not physically gathered] and into a state of frenzy in which attempt at rational argument only spurs the animal spirits" (Schumpeter 1942, p.257). Barber expresses some concerns regarding this point (Barber 1998, pp.585-87).

59. The claim that this is a conscious design to keep citizens away from the political arena is discussed in Part II of *The Betrayal of an Ideal.*

60. Barber (1984), p.272.

61. One should recall that the conception of the individual developed in Chapter 1 went in the opposite direction by taking the individual as socially-oriented and socially-interested.

62. Miller uses this argument to repel the contention that "*the vast majority prefer to leave this task [political decision-making] to professional politicians*" (Miller 1989, p.330, original italics).

63. *The Betrayal of an Ideal* argues that the degrading of the individual citizen's political-moral capacities began with Plato and culminated in Machiavelli and Hobbes.

64. Barber (1984), pp.288-89, italics added. Barber's contention here is similar to Dewey's argument against the exaggeration of the role of expertise in policy making: "It is not necessary that the many should have the knowledge and skill to carry on the needed investigations; what is required is that they have the ability to judge of the bearing of the knowledge supplied by others upon common concerns" (Dewey 1954[1927], p.209).

65. Barber (1984), p.234.

66. The phrase inside the quotation marks is borrowed from Allison M. Jaggar, who argues that in a true multicultural democracy ("composite democracy," as she calls it), cultural differences or "borders" become "'frontiers, spaces of mingling. Instead of dividing they become bridges between other regions, between groups'" (Jaggar 1999, pp.327-28).

67. To those who see cultural pluralism or culturally related or culturally based approaches to politics as divisive, Valadez has an eloquent rebuttal: "What these critics

have failed to note . . . is that cultural conflict and Balkanization are caused primarily not by preservation of and adherence to cultural traditions, but by historical and existing discrimination and oppression of ethnocultural groups and most importantly, by the lack of just sociopolitical institutions for equitable democratic representation and the negotiation of cultural group interests and needs. Identifying with one's cultural tradition in and of itself does not necessarily lead to social divisiveness" (Valadez 2001, pp.19-20). Jaggar presents a similar argument: "Social stability and civil solidarity are not necessarily undermined by cultural diversity: the real threats come instead from cultural insensitivity, injustice, domination and imperialism" (Jaggar 1999, p.327).

68. Save the questions of Quebec and the aboriginal people (which are primarily questions of federalism vs. nationalism), and to a lesser extent the problem of racism (directed primarily against the people of African origin), Canada is a true case of success in multiculturalism. The general platform that could serve as the framework for the public forum in question in the Canadian case would be the Multiculturalism Policy of Canada, declared in 1971, and its manifestation as law in the Canadian Multiculturalism Act of July 1988. These, in combination with the Canadian Charter of Rights and Freedoms, serve as the bedrock for a grand social-political culture wherein different cultures come together to form a vast union of diverse cultures. What is admirable about the Canadian model of multiculturalism, in comparison with the model in the United States, is that with the exception of the Charter, the terms of integration into this grand culture are adjustable. Of course, the Canadian multiculturalism has its problems and has come under fierce attacks in Canada during the last few years. However, as Will Kymlicka has persuasively argued in his *Finding Our Way* (1998), the critics of Canadian multiculturalism are short on empirical evidence, as well as being "simply uninformed about the consequences of the policy" (ibid., p.16). Kymlicka has also argued that the presence of uncertainty and suspicion about the apparent anti-liberal dangers hidden in multiculturalism (e.g., the diluting of some civil liberties in some cultural communities), and the overall opposition to the idea among Canadians, should be attributed to the failure of the federal government to convey to Canadians that multiculturalism has "'non-negotiable' requirements," and that it does not "undermine respect for liberal-democratic values . . . [or] . . . institutional integration" (ibid., pp.65-70). To remedy this oversight, Kymlicka proposes a "public debate over the limits of multiculturalism" (ibid., pp.68-70).

69. Iris Marion Young has made this argument (Young 1995, p.209). Her argument was cited earlier in this chapter.

70. As was argued in Chapter 3, the main positive feature of having diverse political opinions is that it would safeguard against the tyranny of the majority and would also prevent demagogues from rising to the top.

71. This question was discussed in the case of the example of gay-lesbian marriage in Chapter 2.

72. The case of Athenian democracy is discussed in Part I of *The Betrayal of an Ideal*.

73. Quoted by Femia (1996), p.389.

74. Face-to-face discussions have a number of other shortcomings, some of which were discussed in Chapter 3 in this volume and Part III of *The Betrayal of an Ideal*.

75. A reference to the famous utterance of Oscar Wilde (Cf. Levine 1993, p.92).

76. Two comments are in order here. First, on the question of empowering "introverts" through electronic technologies, McLean argues that "introverted people have opinions too, and in a democracy they should have an equal right to have them heard. The new information technology could put them on an equal footing for the first time" (McLean

1989, p.94). Second, apropos of the "interactivity," Budge has argued that debates held via the use of electronic technology can also be "fully interactive" (Budge 1996, p.28).

77. It was argued in Chapter 3 that the existence of a politically educated and responsible citizen body, the presence of a diversified field of political views and interests, and the situation wherein there exist multiple loci of power (distributed among citizens, individual guardians and representatives, and their assemblies) would prevent demagoguery from going overboard. As an added protection and guard against the "whimsical" tendencies of the people, the constitution of the Realistic Democratic Utopia could be amended to require that each adopted policy option be reaffirmed after a period of six or nine months. This is similar to Barber's idea of two-stage voting, or "second reading," as he calls it (Barber 1984, p.288). If an adopted policy option fails to be reaffirmed, then the original decision would be regarded as nullified. This would be taken as an indication that, by being whimsical on the issue, the public has disqualified itself from making a decision on it; and that decision on that particular issue should be made by the assemblies of the guardians and the representatives. For the reasons of efficiency and preventing the over-politicization of society, the reaffirming vote would be a re-run of the run-off vote between the two options that made it to the run-off phase in the original voting occasion. Moreover, for the same set of reasons, the reaffirmation vote would not be preceded by extensive public deliberations. Lastly, in making the final decision (when the public disqualifies itself by being whimsical), the guardians and representatives would vote to choose one of the two policy options that have made it to the run-off vote.

78. Here the phrase "overlapping consensus," and the general idea behind it, are borrowed from Rawls (Rawls 1993, p.147-48). Moreover, the fundamental idea behind the notion of "elastic and publicly accepted conceptions of the 'good life' and the 'good society'" is similar to the idea behind Rawls' "reasonable comprehensive doctrines," not all of which would necessarily be liberal (Rawls 1993, pp.58-66 and pp.xxxix-xl). What makes this vast pluralistic community a "nation," in addition to what was stated above, is its citizens' common points of reference, shared territory, common experiences, and a conception of a common future. It goes without saying that each society has its own particular family of conceptions about "good society" and "good life." These conceptions are rarely formulated in terms of concrete statements and theses. Rather, they are manifest in manifold societal spheres, entities, and practices. The hegemonic conceptions in each family shape the building blocks of constitutions and the intellectual-cultural-moral life. They are affirmed in laws, as well as in philosophy, art, literature, and other intellectual products of each society writ-large. The following is an attempt toward formulating a framework for developing an "overlapping consensus" from elastic and generally accepted conceptions of the "good life" and "good society" that would be congruent with the political structure of direct-deliberative *e*-democracy (or the Realistic Democratic Utopia outlined in Chapter 3). A good life is a life that attends, nurtures, and develops both the natural and social sides of Man's being. In the sense that it attends to her natural needs, it takes Man as an individual, individualistic, egoistic, self-interested, and spiritual as well. However, insofar as it takes Man as social, it needs to consider the social, communal, moral (personal, social, and political) aspects of his being and development. These latter considerations go beyond, and could come into a conflict with, attending to Man's natural and individualistic needs and expectations. And here, the social dimension of the "good life" comes into play, which is tied to a positive conception of human development. A "good life" is a life that is directed toward the realization of a set of morally valuable ends that could be regarded as "universal," and as "intuitively" and "objectively" valid. These are the ends that have withstood the test of history and have come to

be recognized by the general consensus of the past and present generations in the Western liberal-democratic societies—if not in most present-day societies—as ends valuable in themselves. These ends, and the enduring values upon which they stand, form the common denominator and core of the aspirations and longings expressed in art, literature, philosophy, and religion; and for these reasons, they could be regarded as "universally" desirable and worthy values to pursue. Reverence and compassion for life; striving for peace, harmony, solidarity, and social stability; respect for individual autonomy; belief in the dignity and worth of each individual, in fairness and equality, in the amelioration of life's suffering, in pacification and improvement of the conditions of human existence, and in the positive development of the human individual (the artistic, intellectual, moral, and physical) would be among these ends. (In addition to being historically developed, as was suggested in note 16 in Chapter 1, one can also argue that these ends are also derivable in the Rawlsian original position—and independent of any comprehensive moral, philosophical, or religious doctrines.) This broad and elastic conception of the "good life" is based on a positive conception of human development. The conception of the "good society," on the other hand, is an extension of the concept of the "good life." A "good society" is the one that is committed to the principle of the pacification of existence; a society that provides maximum feasible opportunities for all individuals to pursue and attain a good life, and does so on an equitable basis. What reconciles the natural and social aspects of Man, as was discussed in Chapter 1, is her *consciousness*.

79. The assumption of "common or collective identity," as well as the earlier assumption that some "elastic and publicly accepted conceptions of the 'good life' and the 'good society'," and along with them, the assumption that "a shared broad conception of fundamental universal values and ends regarded as worthy" are achievable in present-day America seem quite reasonable in light of what Rawls calls "The Fact of Reasonable Pluralism," "The Fact of Democratic Unity in Diversity," and "The Fact of Public Reason" (Rawls 1999, pp.124-25). Rawls argues that these facts are "confirmed by reflecting on history and political experience" of the liberal-democratic societies (ibid., p.24). Concisely stated, the first fact indicates that "[d]ifferent and irreconcilable comprehensive doctrines will be united in supporting the idea of equal liberty for all doctrines" (ibid.). The second fact indicates that "political and social unity does not require that its citizens be unified by one comprehensive doctrine, religious or nonreligious" (ibid.). Finally, the third fact makes the point that citizens holding different and irreconcilable comprehensive doctrines accept the "idea of the politically reasonable." That is to say, when putting forth their doctrines or introducing political policies derived from them, they "provide sufficient grounds in public reason" for them (ibid., p.125).

80. In this respect, the commitments and responsibilities of citizens in direct-deliberative *e*-democracy would be "self-assumed." Moreover, in this respect, the theory of direct-deliberative *e*-democracy fits Sandel's characterization of a "more modest" or "instrumental" version of republicanism. In this version, "even the liberty to pursue our own ends depends on preserving the freedom of our political community, which depends in turn on the willingness to put the common good above our private interests" (Sandel 1996, p.26)

81. Here, one is reminded of Tocqueville's "doctrine of self-interest well understood" discussed in Chapter 1. Tocqueville believed that the doctrine had to be promoted by enlightening the people: "Enlighten them . . . at any price . . . I see the time approaching when freedom, public peace, and social order itself will not be able to do without enlightenment" (Tocqueville 2000, Vol.2, p.503). In this sense, the guardians will function as the "moralists" and "enlighteners" in direct-deliberative *e*-democracy.

82. To the charge that this response to the question of "Which takes priority over the other, individual or community?" shuns the question, the response would be that this is the whole point. The idea is to dissolve and de-legitimate the question, for the dichotomy of the individual versus community is irresolvable in theory.

83. As was mentioned in note 27 in Chapter 3, the idea of "local talk shop" is borrowed from Barber. A local talk shop is where people socialize, e.g., civic homes, marketplace, and barbershop (Barber 1984, p.268).

84. The idea of "civic home" is also borrowed from Barber (ibid., p.271).

85. Among the excluded institutions are some of the parochial schools and universities, exclusive clubs, and the private media that operates solely on the basis of profit.

86. The phrases in quotation marks are borrowed from Putnam (2000), pp.18-24.

87. This conception of civil society is similar to Habermas' conception insomuch as he distinguishes civil society from the economic society. However, it differs from his in that Habermas tends to reduce civil society to his notion of "public sphere" and through that to a communicative network, e.g., see Habermas (1996), pp.366-67. Though Gramsci is regarded by some (e.g., Simon 1982, pp.69-70) as a forerunner of the idea of distinguishing the economic society from the civil society, Gramsci's writings seem to be mainly suggestive of this separation. As was the case with Hegel and Marx, Gramsci makes a distinction between the "political society," on the one hand, and the "civil society," on the other hand—the latter being "the ensemble of organisms commonly called 'private'" (Gramsci 1971, p.12). Gramsci also maintains that the two constitute an organic unity and that their separation is "merely methodological" (ibid., p.160). He only hints at a division within the civil society when he suggests that in "the most advanced State" civil society can withstand the "incursions" of the economic crisis (ibid., p.235). While Simon's claim that Gramsci believed "a capitalist society . . . [was] . . . composed of three sets of social relations: the relations of production, . . . the coercive relations . . . the state; and all other social relations" is on the mark, his conclusion that Gramsci explicitly regarded the latter set of relations as making up the "civil society" seems to be overdrawn (ibid., p.69).

88. Of course, this statement should not be interpreted as excluding the possibility that one's work or labor could become an object of enjoyment and self-development.

89. Putting this problem within the context of the arguments of *The Betrayal of an Ideal*, this is the second democratic shortcoming of the representative form of government under the liberal-democratic state.

90. Gramsci's understanding of hegemony is much more complex than implied above. Concisely stated, Gramsci sees hegemony in terms of the "supremacy" of the capitalist class and its leadership of the society. For Gramsci, hegemony "signifies a determined system of moral life," an "intellectual and moral leadership" (quoted in Morera 1990, p.164). Yet, hegemony for Gramsci, Morera believes, "is not merely an ethico-political phenomenon. It is, rather, the ethico-political aspect of the historic bloc" that the capitalist class forms with its allies (ibid., 168). Its complexity is tied to the complexity of the historic bloc, and to the economic-corporative interests involved. According to Morera,, Gramsci sees the "source" of hegemony (in the capitalist society) and "its original principle" in "the exploitation first of labour but also of other identifiable groups" (ibid., p.175). Simon, on the other hand, emphasizes the ethical-political aspect of hegemony and the important role played by it in the civil society. This in part has to do with the fact that Simon takes Gramsci's sense of civil society as an "ethical or moral society." "[I]t is in the civil society that the hegemony of the dominant class has been built up by means of political and ideological struggles" (Simon 1982, p.69).

91. The sway of the economic society over the whole society in the United States is maintained through its economic resources and its leveraging of governmental resources, as well as through its influence in educational and research institutions, foundations, and the private media. The economic society also maintains its hegemony by the positions its members occupy as directors and board members of these institutions.

92. Recalling from Chapter 3, electronic townhall meetings and televised call-in political talk shows, directed by guardians, and publicly funded television and radio stations, as well as Internet sites, would be some examples of such uses.

93. Given the ease of voting electronically, one can raise the expectation that the people must be consulted more often than once every two or four years. Recalling from Chapter 3, the Realistic Democratic Utopia would conduct voting three or four times a year.

94. FEFI is a good example in demonstrating how far one can go in his expectation of what an electronic voting system can accomplish.

95. For instance, new technologies of production developed in the eighteenth and nineteenth centuries were instrumental in the transition from feudalism to capitalism.

96. The "question of technology" began with an alarmist tone. "Technique," Jacques Ellul declared in 1954, "has become autonomous" (Ellul 1964, p.14 of the English translation). In 1964, Marcuse sounded the alarm again as he boldly asserted that technology served the interests of domination. "In the medium of technology, culture, politics, and the economy merge into an omnipresent system which swallows up or repulses all alternatives"; and that the technical progress is contained "within the framework of domination" and "[t]echnological rationality has become political rationality" (Marcuse 1964, pp.xv-xvi). Contemporary social-political studies on technology, though still alarmist, sound subtle and employ a language that is only vaguely political for the most part. Now, technology is treated more as a social phenomenon and less as a political one. As a result, the focus of research has shifted to the social and cultural spheres. Neil Postman's cultural approach to the question in his *Technopoly* is a good example here. According to Postman, "the uncontrolled growth of technology destroys the vital sources of our humanity. It creates a culture without a moral foundation. It undermines certain mental processes and social relations that make human life worth living" (Postman 1992, p. xii). Some contemporary critics of technology preoccupy themselves with the questions of the degradation of community and the individuals' loss of autonomy and control over their lives (e.g., Franklin 1990). Others focus on the question of whether, or how, the technologically-induced or conditioned practices and discourses force us to call into question the fundamental philosophical assumptions and the values of the modern world. Andrew Feenberg's *Questioning Technology* (1999) and *Alternative Modernity: The Technological Turn in Philosophy and Social Theory* (1995) are good examples here. The most recent views on the question of technology in America espouse differing shades of "substantive" stands on the question of values inherent in technology, yet they resist falling into the cultural conservatist or fatalist traps of "substantive" theories. At the same time, they work with the "instrumentalist" assumptions that technologies are not autonomous, and can be developed and utilized in accordance with democratic principles that fit within the liberal-democratic conception. These views are not systematically or foundationally political, as is the case with the "critical theory." (See Feenberg (1995) for the classifications of the theories of technology into "substantive," "instrumental," and "critical.") That is to say, they do not seek the solution to the social-political question of philosophy in grand-scale social-political transformations as, for instance, Marcuse did. Rather, they focus on studying specific ethical, cultural, and legal phenomena associated with new

technologies or attempt to study specific technologies. (For example, privacy rights, property rights, and moral responsibilities are among themes of interest in the fields of "computer ethics," "bio-ethics," and "business ethics.") Moreover, they tend to identify the solutions in readjusting specific social-cultural attitudes or altering specific institutions, and doing so in pragmatic ways. These views can be regarded as constituting the Anglo-American approach to the social-political dimension of the question of philosophy. (The continental philosophy of technology, on the other hand, is more concerned with the ontological and social-political aspects of the question in large-scale societal contexts and its philosophical roots can be traced to Heidegger and Marx.) Books by Albert Borgmann, Don Ihde, and Langdon Winner are good examples of works that are philosophical in nature but only implicitly or minimally political. Ihde is mainly concerned with the ontological aspect of the problem and the way we make sense of our own existence and the world through the lens of technology and within its medium. Borgmann is interested in studying the relationships between things and signs, and lately, between reality and information, and the nature of the latter (Borgmann 1999). Winner regards technologies as having "political properties," in that they structure our lives in very much the same way the legislations do. In general, he raises more questions than he answers (Winner 1986). Wood works with Henry David Thoreau's observation that "men have become the tools of their tools" (Wood 2003, p.87). A quick survey of some of the recent works that explicitly deal with the issues related to using *e*-technologies for political purposes was provided in the opening pages of Chapter 3, especially in note 6.

97. This disclaimer aside, given that the theory of direct-deliberative *e*-democracy utilizes the latest *e*- technologies for democratic purposes, it has to address a political question often raised in the field of the philosophy of technology, i.e., it has to state whether it regards these technologies as "value-laden" or as "neutral" on the question of democracy. In addressing this question, one can argue that in light of the fact that the theory of direct-deliberative *e*-democracy does not regard the democratic utility it draws out of these technologies as the externalization of some democratic properties or features inherent in them, but only as a product that results from the way they would be used in the Realistic Democratic Utopia, it places itself within the camp of what Andrew Feenberg calls the "instrumentalist theory" in the philosophy of technology. Moreover, in the instrumentalist camp, the theory of technology that seems to be most compatible with the sort of use the theory of direct-deliberative *e*-democracy makes of technologies is Habermas' version. Habermas, according to Andrew Feenberg, "treats technology as a general form of action that responds to the generic human interest in control. As such, it transcends particular political interests and is politically neutral in itself. Value controversy, and hence politics, belongs to the communicative sphere on which social life depends. Technology only acquires a political bias when it invades the communicative sphere. This is the 'technization of the lifeworld'. It is reversible through reasserting the role of communication" (Feenberg 1999, p.7). As will be argued in what follows, the underlying principles of Direct-Deliberative *e*-Democracy, and the purpose to which it intends to utilize *e*-technologies, could help reverse the adverse political effects that result from their current mode of utilization.

98. The arguments presented in Richard E. Sclove's *Democracy and Technology* and Ursula Franklin's *Real World of Technology* are two prime examples here. Sclove works with the fundamental premise that, since "technologies are an important species of social structure, it follows that technological design and practice should be democratized" (Sclove 1995, p.27). Franklin, on the other hand, argues that "prescriptive" technologies (i.e., technologies that break the production process at the workplace into a series of in-

dependently executable tasks) transfer the control of the production process to the manager (Franklin 1990, e.g., p.23). This seemingly small and efficiency-related undertaking at the micro economic level, Franklin contends, has grave social and political implications for society at the macro levels.

99. The contentions listed here follow in the footsteps of those presented by Ursula Franklin. Franklin argues that these anti-democratic consequences are directly linked to the utilization of what she calls "prescriptive" technologies (Franklin 1990, e.g., pp.23-25, pp.37-40, pp.45-48, p.55, and p.116). The term "will to reason" is borrowed from Wood (2003), p.140. See note 6 in Chapter 3 for Wood's argument.

100. In contrast to this conceptualization of the mode of governing in direct-deliberative *e*-democracy, one can argue that the liberal-democratic model (pure representative government) can be conceptualized as a technology of the external control and management of the people.

101. The problem of "digital divide" is widely discussed in the literature. For instance, see discussions in Mossberger *et al.* (2003), Wilhelm (2000), and Norris (2001). Mossberger *et al.* contend that the depth of the problem of the "digital divide" is not fully understood or appreciated by researchers and policy makers. They break up the problem into three categories of "access divide," "skill divide," and "economic opportunity divide," all of which result in the "democratic divide." Wilhelm argues that the problem of "digital divide" is not an insurmountable one and propose providing online services, computers, and training to those who cannot afford digital instruments, e.g., the case of the "Phoenix at Your Fingertip" project. Norris's conclusion is that "[d]igital politics . . . has the potential to amplify the voice of smaller and less well-resourced insurgents and challengers, whether parties, groups, or agencies" (Norris 2001, p.239).

102. See Mossberger *et al.* (2003), pp.38-47 for a detailed discussion of the problem of "skill divide." See also Wilhelm's discussion of this problem and his proposed solution in Wilhelm (2000), e.g., the case of the "Phoenix at Your Fingertip" project.

103. This classification of criticisms and concerns related to utilizing *e*-technologies for democratic purposes is by no means exhaustive. There are a number of other potential questions and problems that skeptics often bring to light. In many respects, these problems have more to do with questioning the practicality of the idea of *e*-democracy itself than with the *e*-voting aspect of the problem. Among these, one should mention the social-cultural questions that relate to concerns about the citizens' diminishing civic capabilities (e.g., their growing political illiteracy and apathy, and their ever-shrinking attention spans) and the problem of information overload and the citizens' inability to put in context, or make sense of, information that bombard them on a daily basis. (See Wood (2003), pp.102-03, pp.137-45, and pp.167-74 for a statement of the problem.) The skeptic reader can also point to the political and legal difficulties that could be encountered in attempting to harness the private media of mass communication for the purpose of putting their resources in the service of civil society and democracy. The same sort of skepticism can be expressed regarding the potential difficulties in attempting to prevent the economic society from infiltrating an ideal *e*-democracy. All these "realities," as the potential arguments could go, make the idea of direct-deliberative *e*-democracy an unrealistic proposition. In responding to this argument, one should begin with contending that these problems should not be considered as the inevitable consequences of *e*-technologies, but only as the adverse effects of the ways in which the present society is utilizing them; and that the civil-society-centered system of governing in the direct-deliberative *e*-democracy would remedy these social-political ills brought about by the market-centered modes of utilizing technologies that primarily intend to serve the interests of

the economic society. Now, one can respond that these prevailing "realities," admittedly, throw hurdles in the way of achieving the Realistic Democratic Utopia; however, they do not make it unrealistic. The skeptics should reread pp.104-05 and 142-46 above for a discussion concerning how realistic the idea of the Realistic Democratic Utopia is, and how or whether it can be realized. Under the least favorable circumstances, the Realistic Democratic Utopia should be regarded as a vision of what is possible and an ideal that can serve as a guiding thread in working toward a more democratic society.

104. In the months leading up to the November 2004 elections, many computer security experts and non-profit organizations raised concerns over the fact that the majority of the electronic voting machines being readied for the election (with the exception of the ones in the state of Nevada) did not leave paper trails, and for this reason, their software could easily be tampered with in order to rig the voting results. The reader should consult Dugger (2004) for a detailed discussion of the concerns that surrounded the electronic voting machines in the pre-election period. The two most widely known expert critics of the paper-trail-less voting machines were—and continue to be—Aviel D. Rubin and David Dill, computer scientists and professors at Johns Hopkins University and Stanford University, respectively. David Dill operates the website <http://www.verifiedvoting.org>. There were other Internet sites which sounded alarmist before the elections, and are currently studying the performance of the electronic voting machines on the election day. Among these, one should mention <http://www.blackboxvoting.org>, <http://www.countthevote.org>, and <http://www.notablesoftware.org>. The dire predictions of widespread problems with electronic voting machines did not materialize in the November 2004 elections, despite the fact that there were numerous minor incidents of hardware and software failures, machine malfunctions, and vote-counting errors. (Two failures were widely reported in the media. In the first case, electronic voting machines lost more than 4,500 votes in North Carolina. The other failure took place in Ohio, where the electronic voting machines gave George W. Bush almost 4,000 phantom votes. However, given that the State of Ohio utilized these machines only in a small number of counties, they did not figure prominently in the controversy over vote-counting irregularities and other problems that were encountered in Ohio on Election Day.) As to the fears about major software failures, software tampering, and hence electronic vote rigging, by many indications, it seems that they too failed to materialize. Given that most electronic voting machines did not leave paper trails, if there were rigging, in the words of Aviel D. Rubin, "there's no way to know now anyway". (See "E-Voting Expert Rubin Talks to N-L" <http://www.jhunewsletter.com/vnews>, (6 December 2004)). See also note 106 below. Finally, electronic voting machines are increasingly becoming standard voting tools on a global scale. They were used successfully in the Indian national elections in the Spring of 2004 on a massive scale. They were also utilized extensively in the Venezuelan and Brazilian elections in the Summer and Autumn of 2004.

105. Portable electronic voting machines are already in use. In the November 2004 elections, Florida voting precincts made these machines available to voters who had ambulatory problems.

106. Most of the vehement critics of electronic voting machines find the "paper trail" and random checking as satisfactory solutions to the security question. Both David Dill and Aviel D. Rubin regard the "paper trail" approach as a satisfactory short-term remedy for the security shortcomings of the existing voting machines. (See note 104 above. See also "Computer Expert: E-Voting Systems Flawed" (<http://www.cnn.com>, (6 May 2004)). Another example here is the solution proposed by Congressman Dennis Kucinich, a Democratic candidate in the 2004 U.S. presidential primaries. Kucinich re-

garded the use of electronic voting machines as "dangerous to democracy," yet expressed willingness to accept them provided that they produced "voter-verified paper record for use in manual audits and recounts," and were subjected to "mandatory surprise recounts in 0.5% of domestic . . . and . . . overseas jurisdictions." As an added measure of security, he proposed "ban[ning] the use of undisclosed software and wireless communications devices in voting systems." Moreover, he proposed that special "electronic voting systems be provided for persons with disabilities"—Dennis Kucinich's website (<http://www.kucunich.us/issues/e-voting.php> (7 March 2004)).

107. See Mossberger *et al.* (2003), pp.116-38 for an opposite view and their recommendations for dealing with the problem.

108. See Mossberger *et al.* (2003) for some public policy recommendations for addressing the problems of the "skill divide," "digital divide," and ultimately, the "democratic divide." Direct-deliberative *e*-democracy would need to take much aggressive and bolder attitudes toward overcoming these problems than the ones suggested by the authors.

Conclusion

Predisposed to the view that the latest innovations in *e*-technologies have much to contribute to overcoming some of the logistical difficulties posed by the size and complexities of the modern nation-state, this volume revisited the neglected idea of *direct democracy*. Moreover, armed with the contention that the electronic communication networks, media, and devices brought about by the recent technological innovations have the capacity to give real and practical expressions to the idea of the ordinary citizens' direct participation in political decision-making, the book conceptualized the original idea of democracy (rule by the people) as the "conception of the political empowerment of the citizen." This conceptualization, it was argued, not only remains true to the ideals represented by the original idea of democracy (and embraces the full breadth of its moral substance), but also incorporates into itself, in a coherent fashion, some of the greatest theoretical achievements relevant to the question of democracy that have come out of the tradition of Western political thought.

As an integral part of arriving at this conceptualization, the moral substance of the original idea of democracy had to be rescued from the perversions and distortions it has suffered throughout the history. This constituted the task of *Democracy as the Political Empowerment of the People: The Betrayal of an Ideal*. The companion volume restored to democracy its main and most pressing ideal, namely, the ideal that democracy is about citizens being the sovereign in the state and exercising their sovereignty *directly—that citizens themselves should be the actual decision-makers;* or that they should have the power to ratify or reject, "in person," the decisions made on their behalf by their appointees.[1]

With this ideal rescued in the companion volume, the present volume attempted to reformulate the original idea within the theoretical-cultural framework of present-day American liberal democracy. This reformulation began on a philosophical plane with an attempt to conceptualize the original idea of democracy as the political empowerment of the citizen. In the course of further elaborations, the new conceptualization was fashioned into a theory of democracy that was referred to as the "theory of direct-deliberative *e*-democracy." The "conception of democracy as the political empowerment of the citizen" was originally expressed in terms of two *ideal* principles of the "political sovereignty" and "social autonomy" of the individual citizen. In *real* terms, the citizen's political power was expressed as the power to "fully express" her political wills *and* to "fully integrate" these expressions into the collective policy- and decision-making process.

Focusing primarily on the principle of the "political sovereignty" of the citizen, and thus considering the question of practicing democracy primarily at the macro levels of state and national governments, the next step in developing the theory of direct-deliberative *e*-democracy was to consider how these two powers of the individual citizen could be given actual or *practical* expression in present-day American liberal democracy.[2] This consideration resulted in a complex collective decision-making scheme that was referred to as FEFI ("Full Expression and Full Integration"). The third step in developing the theory was to construct a "Realistic Democratic Utopia." The Realistic Democratic Utopia served as a theoretical-institutional framework which helped develop an understanding of how a highly democratic society could be conceptualized on the principles of the "conception of democracy as the political empowerment of the citizen" and the collective decision-making scheme of FEFI. Finally, the last chapter of the book attempted to synthesize the new conception of democracy, FEFI, and the Realistic Democratic Utopia into a coherent theoretical whole and thus presented the theory of direct-deliberative *e*-democracy in its fully elaborated form.

The hallmark of the theory of direct-deliberative *e*-democracy is that it rests (as was the case with the "original idea" and the "classical doctrine") on a thick notion of sovereignty. The theory is relentless in the pursuit of the idea that democracy is to be identified primarily with the citizens' *actual*, *direct*, and *continuous* exercise of sovereignty (i.e., the citizens' direct and continuous participation in social decision-making on major issues). On this question, the theory of direct-deliberative *e*-democracy diverges considerably from the liberal-democratic theory which is premised on a limited, indirect, and intermittent-periodic exercise of sovereignty by private citizens that takes place exclusively during the election of representatives. The theory of direct-deliberative *e*-democracy also distances itself from the liberal-democratic theory on the question of democratic legitimacy. It steadfastly holds on to the contention that democratic legitimacy is not primarily about securing the "consent" of the people, nor just about establishing "procedures" or assuring their "fairness," nor just about justifying the power of authority; but, most importantly, it is primarily about the direct and continuous input of citizens into the decision-making process. More than anything, democratic legitimacy is a question of the degree to which individual citizens are empowered to experience political power directly, and on an ongoing basis. This contention rests on the conviction that democracy is about providing and facilitating the highest feasible degree of the direct and actual participation of the citizens in decision-making about the major policies that affect their lives, and the fundamental laws that shape their society.

It is worth stressing again that this notion of democratic legitimacy in the theory of direct-deliberative *e*-democracy does not rest on the claim that the collective wills or decisions produced in direct-deliberative *e*-democracy would necessarily be "true" or "fair"; nor does it rest on the contention that these decisions or collective wills would be "correct" or "right." Rather, it rests primarily on the claim that the theory facilitates the *direct, full, ongoing* (and *equal*) participation of all citizens at most, if not all, levels of political decision-making. It is in this light that, in the theory of direct-deliberative *e*-democracy, concerns and questions regarding the "truth," "fairness," or "correctness" of the collectively made decisions are considered as secondary to the concerns regarding the political-moral questions of facilitating the direct, full, ongoing, and equal participation of all citizens in the political process.

In addressing these secondary questions and concerns, the theory of direct-deliberative *e*-democracy took a two-pronged approach. On the one prong, the theory dealt with these issues and questions through its components of guardianship, public education, and public deliberation. In the institutions of public education and deliberation, and under the directorship of the guardians of the public interests, citizens would subject their opinions and preferences to the scrutiny of factual knowledge, moral-rational considerations, and reasonableness. In this process, the institutional values of society, knowledge, normative considerations, and reasonableness would come to educate the wills of citizens before they are entered into the collective will-formation process. On the other prong of this approach, the theory of direct-deliberative *e*-democracy dealt with these questions by reformulating them (when appropriate) in ways that they would lend themselves to being addressed by technical solutions (i.e., social-institutional and mathematical-technological ones). The technical solutions in question would make it possible to design a collective decision-making scheme that would be regarded by the general consensus of a knowledgeable and politically-educated citizenry as an optimally fair, reasonable, and accurate method for amalgamating and channeling the wills of the citizen body into collective decisions. Thus, in the theory of direct-deliberative *e*-democracy, the democratic legitimacy inherent in the principle of direct, full, ongoing, and equal participation of all citizens in decision- and policy-making is buttressed by two other criteria of legitimacy: first, the "moral-epistemic legitimacy" of collectively made decisions (drawn from the existence and proper functioning of the institutions of guardianship, public education, and public deliberation); second, the legitimacy bestowed upon these decisions by the consensus of an informed and politically-educated citizenry who regards the collective-decision-making scheme as optimally fair, reasonable, and accurate.

Notwithstanding its differences with the liberal-democratic theory on the question of democratic legitimacy, in light of being developed within the theoretical-philosophical framework of present-day American liberal democracy, the theory of direct-deliberative *e*-democracy is characteristically liberal-democratic at its very core. This is clearly evident in the fact that it is from the liberal-democratic tradition that the theory of direct-deliberative *e*-democracy acquires the three components of its constitutionalism, viz., its proceduralism, its unswerving commitment to the individual's rights and liberties, and its strong notion of formal political-legal equalities. On the question of constitutionalism, what distinguishes the theory of direct-deliberative *e*-democracy from the liberal-democratic theory is that the former takes these components a bit further, as it holds them up to the light of the idea that democracy is to be understood as the actual, direct, and ongoing participation of individual citizens in decision-making on matters that have profound impacts on their lives. In so doing, the theory of direct-deliberative *e*-democracy achieves three things: it breathes substance into the liberal-democratic proceduralism; it expands the latter's scope of rights and liberties to also encompass a "positive realm" by politically empowering the individual; and it extends the liberal-democratic commitment to equalities to also include a serious attention to the question of social and economic equalities, albeit not as an absolute prerequisite for democracy, but mainly as a fortifying and bolstering element.[3]

While it acquires its constitutionalism from liberal democracy, the theory of direct-deliberative *e*-democracy develops its appreciation for the epistemic and morally educative values inherent in the idea of public deliberation from the *earlier* theories of deliberative democracy. In agreement with the fundamental intuition that drives these theories, the theory of direct-deliberative *e*-democracy regards the idea of public deliberation as a necessary condition for democracy. However, in disagreement with them, it regards the idea as insufficient in itself, and thus in need of being complemented with the idea of the direct exercise of sovereignty by individual citizens. Without this latter component, the theory of direct-deliberative *e*-democracy contends that democracy falls short of politically empowering the citizen. In the case of the most recent theories of deliberative democracy, the theory of direct-deliberative *e*-democracy holds that, even if the democracies of "public justification" and "public contestation" could be actualized, the best one could hope for under their reign would be a "responsive" government, and not necessarily a "responsible" government or a government "for" the people. As with the liberal-democratic conception, the main problem with the theories of deliberative democracy is that they fall short of affirming that the ultimate guarantor of a responsible government (or a government "for" the people)—as well as the true remedy for the democratic shortcomings of representative government—is a government "by" the people where individual citizens directly participate in performing the legislative and policy-making functions of governing.

* * * * *

Considered in a slightly different context, the main idea here is that, in developing its appreciation for the moral and epistemic values inherent in the idea of public deliberation, the theory of direct-deliberative *e*-democracy moves beyond the theories of deliberative democracy to also appropriate the value of "self-rule" or the direct exercise of sovereignty—a value which is historically associated with the idea of public participation in the tradition of "civic republicanism."[4] However, in light of the jurisprudential limitations it accepts by appropriating the liberal-democratic model of constitutionalism, and also given its appreciation of the depth and breadth of the difficulties the large size and complexities of the modern nation-state pose for the idea of direct rule by the people, the theory of direct-deliberative *e*-democracy embraces the idea of the citizens' self-rule only in the *legislative* sense of the idea (and then primarily for the "major issues" of concern to the public).[5] In this sense, the theory of direct-deliberative *e*-democracy is to be regarded as a "*pragmatic*" theory of *direct democracy* or as a theory of "*constitutional*" *direct democracy*.[6] Furthermore, in the sense that the theory of direct-deliberative *e*-democracy reconciles liberal-democratic constitutionalism with the republican notion of self-rule, it presents itself as the dissolution of the dichotomy between the "liberty of the ancients," on the one hand, and the "liberty of the moderns," on the other hand.[7]

In connecting with the republican tradition in political philosophy, the theory of direct-deliberative *e*-democracy grounds itself upon some of the fundamental ideas in Rousseau's thought. Stated summarily, from Rousseau, the theory of direct-deliberative *e*-democracy acquires its contentions that the direct exercise of popular sovereignty is the primary criterion of democratic legitimacy, and that sovereignty cannot be represented. In agreement with Rousseau, the theory of direct-deliberative *e*-democracy conceptualizes citizens' direct participation in sovereignty in its legislative sense and stresses the educational values inherent in the idea of the direct participation in politics. However, the theory of direct-deliberative *e*-democracy distances itself from Rousseau's "general will" and abandons his aconstitutionalism, especially on the question of protecting the (negative) liberties and rights of the citizens. The theory of direct-deliberative *e*-democracy also parts company with Rousseau on his rejection of the idea of public deliberation and his idealization of the leader as a charismatic (and manipulative) supreme lawgiver.[8]

Moreover, in accord with much of what republicanism, and communitarianism for that matter, stand for, the theory of direct-deliberative *e*-democracy stresses the importance of community and civic virtues, and thus, in agreement with them, emphasizes the need for cultivating civic skills, nurturing community, and protecting its interests. In affirming these, the theory of direct-deliberative *e*-democracy appropriates a fundamental thesis of Plato's *Republic*, namely, the idea that political decision-making ought to be embedded in reason, morality, and virtue.[9] This appropriation goes hand in hand with embracing the closely related idea of guardianship in *The Republic*, albeit in a different sense and for a radically different purpose than Plato's own intention.[10] As was discussed in Chapter 4, the guardians in the theory of direct-deliberative *e*-democracy are conceptualized as representing both the social-moral conscience, and the ideally-high social-political consciousness of the citizenry. Furthermore, guardians are also conceptualized either as being well-versed in matters of social, political, and economic policies, or as having exemplary experiences and knowledge in serving the community, and thus as having the credentials needed to command the respect and trust of the citizens. These credentials would enable guardians to bring knowledge, "truth," reason, wisdom, and integrity both to the practice of politics in the state and to the larger arena of the public political culture. They would accomplish this not only by directly partaking in decision-making (in the state) but, more importantly, by informing and influencing the citizens' self-education and will-formation activities (in the civil society).

Straying for a moment from exploring the republican connections of the theory of direct-deliberative *e*-democracy, it is worth arguing at this point that the guardianship component enables the theory to lay claim to offering a solution to a fundamental philosophical problem that dogged ancient and early modern political thought, viz., the now-forgotten problem of reconciling the egalitarian ideal of governing by the consent of the governed with the equally desirable ideal of governing in accordance with principles of wisdom.[11] Considering the responses of Plato and Rousseau, the two main thinkers who attempted to tackle the problem head-on, it is easy to see why neither could offer a satisfactory solution to the problem within the context of his era. Plato absolutized the principle of wisdom, as he argued for a pervasive rule of wise men, meanwhile resorting to "noble lies" in order to secure the compliance of the populace. Taking the opposite approach, Rousseau's solution granted the official title of the sovereign to the egalitarian principle of the popular consent. However, at the same time, as was the case with Plato's solution, it, too, turned to deception—in the form of the manipulation of the general will—in order to assure that, in the end, it was the rational will of the wise men, and not the popular will, that reigned as the actual sovereign power in the state.[12] Returning to the theory of direct-deliberative *e*-democracy, one can argue that this ancient philosophical problem also lies at the heart of the theory and that the theory can be viewed as an attempt to revive the ancient question. Moreover, one can further argue that the theory poses the question in a characteristically Rousseauean context.[13] What is more, the theory also offers a characteristically Rousseauean response to the problem. It, too, grants the official title of the sovereign to the principle of consent. However, thanks to the guardianship component, its response differs from Rousseau's in that the power the theory bestows on the principle of popular consent is as *genuine* and *actual* as it is official. The success of the theory in providing a satisfactory solution to the problem comes about in two ways: first, through having in place a system of political decision-making that requires the political cooperation of the elites and the private citizens; second, and more importantly, through bringing wisdom and knowledge to the discourse of public political culture—thus facilitating a reconciliation between principles of consent and wisdom on both the political and the moral-intellectual planes.

In returning to the main question of this section, one might ask: What exactly are the republican features of the institution of guardianship in the theory of direct-deliberative *e*-democracy? While the theory of direct-deliberative *e*-democracy grants the citizens a constitutionally-protected rich package of basic (negative) liberties and rights, and places effective political powers at their fingertips (positive liberties), at the same time, it counterbalances these powers with the power of a culture that nurtures civic virtues, and values the importance and relevance of knowledge, reason, normative considerations, and virtue to political decision-making. The theory achieves this counterbalancing by assigning to guardians a powerful educative authority in the civil society. This authority enables the guardians to keep civic concerns and the collective interests of the community present in the discourse of the public political culture and, through that, in the consciousness of the citizens. Furthermore, the constitution of direct-deliberative *e*-democracy and, along with it, the public political culture and its highly respected guardians, impel the citizens to live up to their citizenship responsibilities, and in exercising their political power placed at their fingertips, assign heavier weights to the interests of the community than to their private ones.[14]

Now, in light of these considerations, the theory of direct-deliberative *e*-democracy is in a position to present itself as the resolution of a dichotomous contrast that is often drawn between the "liberal view," on the one hand, and the "republican view," on the other hand. Before making the case for this argument, it would be helpful to present two somewhat different statements of this dichotomy. The following is the statement of the problem as formulated by Sandel:

> On the liberal view, [negative] liberty is defined in opposition to democracy, as a constraint on self-government. I am free insofar as I am a bearer of rights that guarantee my immunity from certain majority decisions. On the republican view, [on the other hand,] liberty is understood as a consequence of self-government. I am free insofar as I am a member of a political community that controls its own fate, and a participant in the decisions that govern its affairs.[15]

For the second statement of the dichotomy, one can turn to Habermas, who poses the problem in terms of the contrast he draws between the republican commitment to the principle of "popular sovereignty" (and the related idea of the self-realization of the individual through serving the common good of an ethical community), on the one hand, and the liberal devotion to the principle of "human rights" (and the individual's autonomy), on the other hand.[16] As Habermas rightly argues, the solitary commitment to the universal moral principles of human rights and individual liberties prevents the liberal view from endorsing the value the republican view sees (and cherishes) in the political will of an ethical community which aims at serving its common good. And, at the same time, the republican view's unswerving commitment to endorsing the popular sovereignty or the will of the community, as liberals correctly argue, tends to treat human rights and individual liberties as "determinations of the prevailing political will."[17]

Because the liberal and republican views take the opposing poles of the individual-community problem as their points of departure, the dichotomy, as posed above, is indissoluble in theory. Now, the claim that the theory of direct-deliberative *e*-democracy poses itself as the resolution of the dichotomy at hand rests on three contentions. First, the theory is premised on *an individuated mode of exerting the sovereignty of the community*. That is to say, the exercise of popular sovereignty or the self-government of the people does not take place through an abstract agency that lays claim to knowing or representing the "general will" or the ethical-political mandate of the community, but rather by the very individuals who constitute the community. The exertion of sovereignty by the community in direct-deliberative *e*-democracy turns out to be nothing other than *the exercise of political autonomy by the individual*.[18] This mode of exerting popular sovereignty shifts the burden of dissolving the individual-community dichotomy from the realm of theory to that of practice, and from philosophy to politics.[19] Second, the self-government or the self-legislation in direct-deliberative *e*-democracy takes place within the confines of a constitution that upholds human rights and individual liberties as a set of universal moral values and, at the same time, places them beyond the reach of democratic decision-making or political will-formation. Finally, in addition to being morally-epistemically constructed (thanks to the institutions of public education, guardianship, and public deliberation), and thus having the good or the generalizable interests of the public as their main object, the laws legislated in direct-deliberative *e*-democracy are also attuned to the private interests of individual citizens.[20]

The last contention above sets the stage for the theory of direct-deliberative *e*-democracy to pose itself also as the dissolution of a "paradox" that some liberal theorists claim to have detected in classical republicanism. This is how Quentin Skinner states the case for classical republicanism, which appears as paradoxical to the liberal view:

> if we wish to maximize our liberty, we must devote ourselves wholeheartedly to a life of public service, [thus] placing the ideal of the common good above all considerations of individual advantage [hence the "paradox" of political liberty].[21]

From the standpoint of the theory of direct-deliberative *e*-democracy, the "paradox of liberty" is born out of the suppositions that public interests (or the common good) stand opposed to the individual's private interests, and that the intent of the law is to interfere with the individual's (negative) liberties or limit their scope. The paradox begins to dissolve once these suppositions are eliminated. In the theory of direct-deliberative *e*-democracy, the individual citizen (as the direct lawgiver or participating lawmaker) is herself the public. Moreover, in direct-deliberative *e*-democracy, the current antithetical relations between the public and the law, on the one hand, and individual liberties, on the other hand, are transformed into dialogical and deliberative relations that eventually produce a compromise or resolution (in each collective decision-making occasion) between the individual's positive liberties, on the one hand, and her negative liberties, on the other hand. Furthermore, as a lawmaker, the individual in direct-deliberative *e*-democracy is not obligated (either morally or legally) to surrender (or sacrifice) her private interests and negative liberties to the public interests. Rather, she is only educated and persuaded by the civic culture to keep her larger and long-term interests (i.e., the public interests) present in mind and to assign heavier weight to them than to her more immediate (private) interests when she exercises her positive liberty in the act of lawmaking.

In closing the discussion of the republican features of the theory of direct-deliberative *e*-democracy, it is relevant to point out here that the theory's approach to the question of democracy has a distinctly Tocquevillean character. Stated concisely, although Tocqueville opted for an essentially liberal-democratic type of political structure, he, as "a liberal of new kind," supplemented this system of government with a set of institutions in the civil society that were intended to counteract both the individualistic (i.e., egoistic and apathetic) and conformist (and thus potentially despotic) tendencies that the liberal-democratic political structures cultivated.[22] The supplementary institutions preferred by Tocqueville ranged from the voluntary associations formed by citizens to the institutions of local self-governments (as well as the educational and apolitical religious institutions), which not only facilitated citizenship participation but also served as educational institutions that cultivated civic mores, attitudes, and habits.[23] Tocqueville believed that these institutions would foster a new culture of citizenship and promote new models of civic participation.[24] As with Tocqueville's approach, the theory of direct-deliberative *e*-democracy also starts with an essentially liberal-democratic form of government, and shares with Tocqueville his insistence that this type of government must be enhanced with a republican type of political culture and a participatory understanding of citizenship.[25]

* * * * *

Now returning to the fundamentally liberal-democratic core of the theory of direct-deliberative *e*-democracy, in addition to what was laid out in Chapter 4, one needs to add that this core was developed within a characteristically Rawlsian context. As is the case with Rawls' theory of justice, and his broader conception of "political liberalism," the theory of direct-deliberative *e*-democracy took shape in the sphere of the *culture* of Western liberal-democratic society. From Rawls, the theory acquired some of its underlying assumptions in Chapters 1 and 4, most notably the ideas of "citizens as free and equal persons," as "reasonable persons"; his conception of "society as a fair system of cooperation"; his "facts" of "reasonable pluralism," "democratic unity in diversity," and "public reason"; and the idea of the "overlapping consensus"—all of which Rawls takes as being implicit in the "public political culture" of present-day American society. The theory of direct-deliberative *e*-democracy also took Rawls' "justice as fairness" as an adequate theory of justice for direct-deliberative *e*-democracy. However, in contrast to a semi-passive, reluctantly social, and apolitical conception of the individual, and unlike a static notion of culture, both of which permeate Rawls' "political liberalism" (and all liberalism in general), the theory of direct-deliberative *e*-democracy posits much potential in both the culture and the citizens of present-day American society and attempts to develop them into socially and politically active and vibrant agents. The theory of direct-deliberative *e*-democracy takes citizens not just as "free and equal *persons*" but also potentially as "free and equal *political actors*." Similarly, it idealizes the American society not just as a "fair system of *cooperation*" but also potentially as a "fair system of *direct participation*." The theory of direct-deliberative *e*-democracy makes the case that the "reasonable persons" and "liberal peoples" of present-day society are capable of doing, and willing to do, much more than just aspire to lofty liberal ideals, have "moral sensibility," be reasonable and tolerant of competing or contrasting conceptions of the good life or the good society, or just agree on a "common ground" on these questions and the question of justice in particular.[26]

In attempting to ground the theory of direct-deliberative *e*-democracy in the tradition of Western political thought, one should also stress that, in addition to being influenced by Rousseau, Plato, and Rawls (and the republican and liberal traditions they represent), the theory is also indebted greatly to C. B. Macpherson and the larger tradition of what one might call "democratic humanism" (most notably represented by J. S. Mill, young Marx, and Dewey) that contributed to shaping his perspective on, and passion for, democracy. The development of the theory of direct-deliberative *e*-democracy became possible in part by relying extensively on Macpherson's developmental models of the individual and democracy, and in part by drawing freely from his analysis of liberal democracy, as well as from his overall project of retrieving the true (i.e., moral) substance of democracy.[27] It was from Macpherson that the theory of direct-deliberative *e*-democracy acquired its conception of what a democratic society ought to accomplish—"to provide the conditions for the full and free development of the essential human capacities of all the members of the society."[28] As a matter of fact, Macpherson's retrieval project was the main motivating force behind undertaking this study. Lastly, from Macpherson, the theory of direct-deliberative *e*-democracy derived its deep-seated conviction that the contemporary (liberal-democratic) conception of democracy holds out a "false promise" to humanity and offers an "impoverished view of life."[29]

Finally, in bringing this study to a close, one should add that, aside from the two democratic shortcomings that haunt the enterprise of liberal democracy (to which the theory of direct-deliberative *e*-democracy poses itself as a solution), liberal democracy also suffers from a closely related problem which it has also inherited from liberalism. As is the case with liberalism, liberal democracy, too, lacks a sense of moral conviction or moral force, or an innate rationality that could help shape an ever-present social consciousness of the need for a broad and overarching (yet non-unitary) conception of the public good that would serve as a guidepost for the society and its citizens. Positing individuals as possessive self-seekers and surrendering the society's political powers (and moral-intellectual-cultural leadership) to the wealthiest among them (and to their world outlooks) are hardly conducive to shaping such a consciousness. If the experience of the past century under the liberal-democratic reign can be regarded as adequate material for extrapolating the *telos* liberal democracy seems to have set for humanity, undoubtedly one is justified in claiming that liberal democracy's *raison d'être* seems to be nothing other than to place society and politics "in the service *of homo economicus*—the solitary seeker of material happiness and bodily security."[30] Moreover, if liberal democracy is working according to some original design, then one is justified in claiming that the entire theoretical construct of liberal democracy must have been erected on a conception of the "half Man" (*homo economicus*) and not the whole Man. Despite liberal democracy's deceptive pretense to uphold the noble principle of the state's moral neutrality, the liberal-democratic society as a whole is asphyxiated by the strong and poisonous presence of an ignoble morality and culture that is largely market-driven and is ultimately devoted to serving the interests of the economic society.[31]

To the eyes beholding it from outside its own moral-intellectual scope, liberal democracy appears as a system of the total mobilization of the whole society for the sake of serving its economy. Seen from outside, liberal democracy's commitment to serving the well-being of its civil society, or to nurturing the civic values that sustain it (or even arguably, its commitment to the protection of individual political liberties), appears as a by-product or extension of its commitment to serving the needs of the market society and securing its moral-intellectual and cultural hegemony. The morality and culture of liberal democracy's marketplace illuminate its "way of life." Life is made to serve the market. While life serves the market, the market feeds life—not just materially, but also morally, intellectually, and spiritually. To this problem of liberal democracy, too, the theory of direct-deliberative *e*-democracy poses itself as a remedy. In the fully realized-phase of direct-deliberative *e*-democracy, it would be the culture and "ideology" of the civil society (the "ideology of life"), and not the "ideology" of the market, that would reign supreme.[32]

The culture of direct-deliberative *e*-democracy, as well as the vision that would guide the movement toward its full realization, would have strong post-material and spiritual dimensions. In contrast to present-day society, whose culture impoverishes life by arresting it in the anxiety- and insecurity-ridden mindset of the marketplace—and impairs its citizens' abilities to see life in its fullness—direct-deliberative *e*-democracy would be a more peaceful society with a more balanced culture. There, the prevailing conception of life, generally seen as struggle for existence, would be transformed into an understanding that would regard living as striving for the development of the whole person and the enrichment of social existence. Self-aggrandization, unbridled materialism, and hyper-individualism would not be the predominant elements of the culture and its norms *par excellence*. The pursuit of artistic, intellectual, and spiritual modes of self-development, as well as self-development in the areas of civic skills and civic life, would be valued as much, if not more than, the pursuit of this-worldly goals and wealth. Life (in all of its civic, social, cultural, artistic, spiritual, and ecological forms) and the pursuit of life-affirming goals would take moral-intellectual and social-cultural priority over market-related or market-driven concerns. Rather than sacrificing life in the service of work (which is the predicament of the majority of the citizens of liberal democracy), or rather than placing civil society in the service of the economic society (the *modus operandi* of liberal democracy), work and the economic society in direct-deliberative *e*-democracy would be placed in the service of life and civil society. While work would be thought of primarily as the means of attaining the individual's livelihood (and secondarily as a subordinate arena for developing her potentials), the economy would be regarded as the sphere of the material production of life and not the end of life itself. It would be deemed as "a means of life and not its despotic master."[33]

Thus, the present mode of life, regarded primarily as material existence and economic activity, would give way to understanding it as primarily civic existence and moral-intellectual activity. The inner longing of the individual would no longer center, as is the case now, on searching for deeper meanings in life that takes place only in those short-lived moments when she manages to free herself from the tyrannical domination of the economic society. In the fully realized stage of direct-deliberative *e*-democracy, the individual would be able to live, in her everyday life, the deeper meaning for which she now longs. She is now a divided being in conflict with herself. She submits herself, voluntarily, yet unwittingly, to the subtle slavery of the material culture, as she throws herself into a frenzied rush in the pursuit of the market-induced goals and desires. However, she never succeeds in reaching an inner satisfaction or a sense of peace she seeks. Not only these goals and desires are essentially unachievable—in that the more she achieves, the more she desires—they are also there in part to petrify her mind and soul, and prevent her from seeing that the true peace she longs for is in herself, and in her community. All she achieves in this frenzied rush is a sense of false security and comfort, as she finds herself in the company of the rest of the restless crowd, as they collectively conform to the norms of the material and consumer culture that shackle their minds and dull their souls. In direct-deliberative *e*-democracy, the driving force behind the individual's life activities would be her longing to develop her whole self. Her life struggles would then center on striking a balance between developing her moral-intellectual and spiritual sides, on the one hand, and attending her this-worldly needs and wants, on the other. Moreover, knowing that her whole self is not entirely in her inner self, she would also seek to strike a balance between her self-focused growth and goals, and her outer needs to seek communion with others in her civic-existence, hence honoring her community-centered and citizenship obligations.

In its fully realized state, direct-deliberative *e*-democracy would be, to borrow the words of the young Marx, a "democratic state" in the sense that it would be "a community of people," "republicans," and "free men" who would have a "sense of self-worth" and work toward the realization of their "highest ends."[34] Democracy then, using Dewey's metaphor, "will come into its own, for democracy is a name for a life of free and enriching communion."[35]

Notes

1. As was argued against Schumpeter's distortions in Chapter 9 of *The Betrayal of an Ideal*, the fundamental idea underlying the democratic theory in the pre-liberal society (or the "classical theory," as he called it), was not that the people must be governed in accordance with their wills, but that *the people themselves must do the governing*. The proof of the sovereignty of the will of the people is not that their will is sovereign in the state, but that the people themselves are the actual sovereigns. Needless to say, once in the position of governing or sovereignty, the people would have no will to govern in accordance with, except the will that they agree upon among themselves (or compose collectively) as their own will.

2. As was mentioned in the closing pages of Chapter 1, the primary concern of this book was the question of democracy at macro levels. Despite this focus, as was contended on several occasions in Chapters 2-4, some of the arguments and discussions presented in these chapters were also relevant to the discussion of the question of democracy at the micro levels.

3. By "breathing substance" into procedure, it is meant that the rules and procedures in the theory of direct-deliberative *e*-democracy are understood as more than just a set of

neutral or formal arrangements or methods for conducting the business of governing (mainly the decision-making aspects); that the substance of the business of governing is to affirm the value inherent in the ideal of the citizens' direct participation in governing—which is ultimately directed toward serving the society's *raison d'être*: the fullest feasible (positive) development of the individual. In this sense, the theory of direct-deliberative *e*-democracy is indebted to theories of participatory democracy and the emphasis they placed on the moral content of democracy (i.e., self-development of the individual). Moreover, the phrase "positive realm" is intended to denote the realm of positive freedoms in the two senses discussed in Chapter 4. Finally, as was seen in Chapter 1, Rawls' principles of justice were regarded as an adequate basis for stressing the importance the theory of direct-deliberative *e*-democracy places on the question of social and economic equalities.

4. By "civic republicanism" here is meant the tradition in political thought currently represented by Michael Sandel. In this tradition, as Sandel puts it, "liberty depends on sharing in self-government . . . means deliberating with fellow citizens about the common good and helping to shape the destiny of the political community. . . . [It] requires more than the capacity to choose . . . requires knowledge of the public affair, a concern for the whole . . . moral bond . . . requires that citizens possess . . . civic virtues . . . requires a politics that cultivates in citizens the qualities of character self-government requires" (Sandel 1996, pp.5-6). One should add that, although the direct-deliberative *e*-democratic notion of "self-rule" is not as thick as Sandel's, it is much thicker than the notions in what Quentin Skinner and Rawls call "classical republicanism"—or what Philip Pettit calls "Roman" or "neo-Roman republicanism." According to Pettit, Sandel's conception of republican liberty is Aristotelian in nature, and despite what Sandel himself believes, his conception is different from the one that has dominated republican political thought since the time of Machiavelli (Pettit 1998, p.49). Pettit is of the view that the conception of liberty that is dominant in this tradition is not the idea of "democratic participation," as Sandel maintains, but rather the idea that liberty requires the "absence of dependency" or "[the absence of] domination by any other" (ibid.). In this sense, as Pettit argues, the republican conception of liberty is different from both the liberal conception ("liberty as noninterference" or negative freedom), and the Aristotelian one as described above by Sandel (ibid.). According to Pettit, the republican conception of liberty as "nondomination" is rooted in the practice of the Roman Republic (rule of law, election to office, term limits in the office, etc.), Ciceronian thought, and in "neo-roman republicanism" (e.g. Machiavelli and James Harrington's) (ibid.). (See Pettit 2002, pp.339-41, Quentin Skinner 1992, pp.216-17, Cunningham 2002, pp.55-59, Rawls 1993, pp.205-06, and Habermas 1996, pp.296-301 for short discussions on the differing views of republicanism. See also Pettit (1997) for a complete treatment of the subject.)

5. Three comments are in order here. First, as is the case with "civic republicanism," the theory of direct-deliberative *e*-democracy understands the idea of "self-rule," or direct participation in decision-making, as "positive liberties" of the citizens, but only in their two senses as described by Macpherson—see Chapter 4. Second, one should be quick to add that, unlike the case of the republican tradition, in which "positive liberties" exhaust the realm of the individual's "political liberties and rights," in the theory of direct-deliberative *e*-democracy, political liberties include both positive and negative liberties. Finally, one should also add that the idea of the citizens' direct participation in performing the executive and judicial tasks of governing (beyond the jury system) is problematic in the modern nation-state. The first problem here has to do with the danger in arbitrariness in having these tasks performed by individual citizens—which would come into

conflict with the procedural requirements. Moreover, there are also the daunting logistical problems in arranging such participation. Third, the high degree of specialization needed in performing some executive or judiciary tasks makes the Athenian idea of full self-rule (legislative, executive, and judiciary) impractical in the modern nation-state.

6. In fact, the pragmatism of the theory of direct-deliberative *e*-democracy has three components. First, it recognizes that direct self-rule in its executive or judiciary senses is not a practicable idea in the modern nation-state. Second, similar to Cunningham's notion of "democratic pragmatism," the theory of direct-deliberative *e*-democracy "draw[s] on work[s] from whatever political perspectives are found useful" and integrates into itself elements from other theories. (See Cunningham 2002, pp.142-49 and Cunningham 1987, p.13.) Lastly, the theory of direct-deliberative *e*-democracy is pragmatic in that it attempts to reformulate some of the theoretical problems encountered in democratic theory in ways that they would lend themselves to being addressed by technical solutions.

7. This dichotomy was formulated by Benjamin Constant. According to Constant, "among the ancients the individual, almost always sovereign in public affairs, was slave in all his private relations . . . as a subject of the collective body he could himself be deprived of his status, stripped of his [citizenship] privileges, banished, put to death, by the discretionary will of the whole to which he belonged. . . . Among the moderns, on the contrary, the individual, independent in his private life, is, even in the freest states, sovereign only in appearance. His sovereignty is restricted and almost always suspended"—from his address to the *Athenee Royale* in 1819 (Constant 1988, pp.311-12). "Individual liberty," according to Constant, "is the true modern liberty. Political liberty is its guarantee, consequently political liberty is indispensable" (ibid., p.323). As was stated in note 5 in the Concluding Remarks to Part I of *The Betrayal of an Ideal*, some (particularly Quentin Skinner and Philip Pettit) have argued that the dichotomous contrast posited by Constant between the liberties of the ancients and the liberties of the moderns is somewhat exaggerated (e.g., Pettit 1997, p.271).

8. See Chapter 3 of *The Betrayal of an Ideal* for a discussion of the problems with Rousseau's "general will," his rejection of deliberations, and his idealization of the supreme lawgiver. See note 71 in Chapter 1 of this volume for a summary of this discussion.

9. As was noted in note 33 in Chapter 3 of this volume, each of the terms "reason," "morality," and "virtue" was used in a broad sense. "Reason" is used to indicate expert knowledge (i.e., social-scientific (and to a lesser extent natural-scientific) knowledge, know-how, and modes of reasoning) relevant to social-political decision-making. The term "morality" stood for the importance of normative considerations or moral thinking in political decision-making. Finally, "virtue" underscored the importance and the need for truth, integrity, and sincerity in politics.

10. The differences between the Platonic idea of guardianship and the one developed by the theory of direct-deliberative *e*-democracy were discussed in Chapter 4. While on the subject of guardianship, it seems appropriate to advance the following contention against the liberal-democratic form of representative government. This form of government, one can argue, is but a perverse form of guardianship, i.e., a form of guardianship in which demos elect their guardians (or rather pseudo-guardians—i.e., the representatives). These guardians are elected more on the basis of their political acumen and skills than their knowledge, virtue, and moral understanding of the issues. Once in a position of power, these pseudo-guardians rule the people without their direct participation in the process. The concept of guardianship in the theory of direct-deliberative *e*-democracy is

both superior to and more democratic than this liberal-democratic perversion of the concept.

11. This statement of the problem is a slightly altered version of the formulation offered by Arthur M. Melzer: "the problem of reconciling the egalitarian principle of *consent* and the elitist principle of *wisdom*" (Melzer 1990, p.240, italics added). Melzer also poses the problem as: "How is one to reconcile the authority of reason and of one's own will, the claim of the good and of the self, the value of rational order and of subjective freedom, the need for wise rule and for self-rule" (ibid., pp.240-41). Melzer revisits this ancient problem within the context of discussing Rousseau's response to the question (ibid., pp. 240-41).

12. This interpretation of the Platonic and Rousseauean responses to the ancient problem draws on Melzer's formulation of the problem (Melzer 1990, pp.240-41). As was argued in notes 74 and 76 in Chapter 3 of *The Betrayal of an Ideal*, the reason Rousseau turned to religious deceit was that he lacked confidence in the cognitive abilities of ordinary people. The same also holds for Plato, for he also lacked faith in the intellectual abilities and moral competence of the common people. Moreover, given that their societies lacked adequate levels of material and technological developments (as well as favorable political-legal structures), informing the consent of the people and bringing wisdom to the public culture could not have appeared to Plato and Rousseau as realistic solutions to the problem.

13. As Melzer points out, the modern political thought has evaded this ancient problem directly and instead has "reduced the political problem to the tension within freedom itself . . . freedom versus 'authority'" (ibid., p.241). This is reminiscent of Levine's argument to the effect that the contemporary political thought has neglected the fundamental questions of the "nature" of sovereignty, authority, and political obligation, and instead has devoted its attention to peripheral questions such as the "limits" of political authority. (Levine 1981, p.13).

14. Here one is reminded of Tocqueville's "doctrine of self-interest well understood" discussed in Chapter 1.

15. Sandel (1996), pp.25-26. Sandel also expresses the contrast between the two in terms of how the liberal view and the republican theory understand the "relation of the right to the good" (ibid., p.25). As was argued above, the theory of direct-deliberative *e*-democracy satisfies a fundamental requirement of the republican theory, viz., the idea of "self-rule." One can contend that the theory of direct-deliberative *e*-democracy fits, albeit not fully, Sandel's characterization of a "more modest" or "instrumental" version of republicanism. In this version, "even the liberty to pursue our own ends depends on preserving the freedom of our political community, which depends in turn on the willingness to put the common good above our private interests" (Sandel 1996, p.26). Moreover, one can also argue that the theory of direct-deliberative *e*-democracy, though weaker than Sandel's own, is much stronger than what Rawls calls "classical republicanism" (in contradistinction to what he calls "civic humanism") in that civic virtues and the citizens' participation in the theory are not posited as means for the protection of liberal rights or freedoms, but rather as means for the self-development of citizens. See Rawls (1993), pp.205-06 and Cunningham (2002), p.57.

16. Habermas (1996), pp.348-52. Here Habermas takes Rousseau and Kant as representing the republican and liberal views, respectively.

17. The phrase inside the quotation marks is borrowed from Habermas (ibid., p.350).

18. Here the phrase "political autonomy" is understood as "positive liberty" in the two senses of the term defined by Macpherson in Chapter 4.

19. Habermas, too, believes that the resolution of the problem of the "liberal view" versus the "republican view" is to be sought *in practice* or, as he puts it, "in the normative content of a mode of exercising political autonomy" (Habermas 1996, p.354). For Habermas, these are but "the procedures for a discursive process of opinion- and will-formation"—i.e., deliberative democracy (ibid.). See Chapter 13 of *The Betrayal of an Ideal* for a short discussion of Habermas' version of deliberative democracy.

20. This argument in part rests on the earlier contention that the collectively-made decisions are those outcomes that are produced by a fair and reasonable scheme of composing the collective wills out of the *fully* expressed wills of the individual citizens.

21. Skinner (1992) p.217. Skinner adds that classical republicanism's preoccupation with the common good is not so much concerned with maximizing individual liberties but with "upholding the ideal of 'free government'" (ibid., p.220). In order to delegitimate the paradox, Skinner goes on to argue that "the classical republican writers never doubt that the majority of citizens in any polity can safely be assumed to have it as their fundamental desire to lead a life of personal liberty" (ibid.).

22. A brief discussion of how, in Tocqueville's view, liberal-democratic structures cultivated these tendencies was presented in note 43 in Chapter 8 of *The Betrayal of an Ideal.*

23. Tocqueville was an enthusiastic supporter of the idea of educating citizens on matters of public interests. As was discussed in Chapter 1, he saw the "doctrine of self-interest well understood" as the focal point of this education. "Enlighten them . . . at any price . . . I see the time approaching when freedom, public peace, and social order itself will not be able to do without enlightenment" (Tocqueville 2000, Vol.2, p.503). In Chapters 8 and 9 of Volume II of his *Democracy in America*—(titled "How the Americans Combat Individualism by the Doctrine of Self-interest Well Understood" and "How the Americans Apply the Doctrine of Self-interest Well Understood in the Matter of Religion," respectively)—Tocqueville comes across as a supporter of the idea of having "moralists," "preachers," and "philosophers" actively participate in the public education of the people (ibid., pp.500-06). In one sense, the guardians in the theory of direct-deliberative *e*-democracy are conceptualized as functioning in ways similar to Tocquevillean "moralists," "philosophers," and enlighteners.

24. Unlike Constant and other "restoration liberals" of his time who saw the problem of their era in terms of contrasting, and eventually choosing between, the realities of modernity and the nostalgia for antiquity—(and who in the end opted for constitutional liberties while abandoning the nostalgia for the "liberties of the ancients" to the dustbin of history)—Tocqueville sought, as Rousseau did, to revive the past and use an idealized image of it as an indictment against the present. Tocqueville's own formulation of the problem of his era did not revolve around choosing between the past and present, but rather around *synthesizing* their most compelling features. In doing so, Tocqueville contrasted the "aristocratic" impulses, which he associated with liberal modernity (individual liberties, "impulse to greatness," etc.) with the democratic inspirations represented by democracy (equality and self-rule). Despite the fact that by his own admission, Tocqueville was "by instinct an aristocrat" and admired the aristocratic "impulse to greatness," love for liberty, and respect for the law; and notwithstanding the fact that he feared and despised democracy for its negative tendencies (anti-aristocratic and tyrannical tendencies, as well as its tendency toward mediocrity); he saw democracy as the solution to the problem of the modern era. Tocqueville, as "a liberal of new kind," attempted to reconcile liberal-aristocratic ideals and inspirations (and the liberal political institutions they would give rise to) with democracy. Tocqueville believed that the destructive tendencies of democracy could be kept in

check by developing a new understanding of citizenship and new models for citizen participation, such as those achieved by Americans (e.g., an apolitical religion, the prevalence of "free *moeurs*," the conception of "self-interest well understood," "decentralized administration," participation in free and secondary associations and in local governments). The interpretation of Tocqueville's answer to the question of democracy offered here is drawn from Neidleman's discussion of the Rousseauean influences in Tocqueville's thought (Neidleman 2001, pp.147-63). Tocqueville's ambivalence about democracy is vividly present in Vol.1, Part II, Chapters 5 and 6 of *Democracy in America*, pp.219-22 of Tocqueville (2000) where he compares democracy with aristocracy—see also p.189 and pp.234-35.

25. Despite this similarity, there remains a major difference between the theory of direct-deliberative *e*-democracy and Tocqueville's approach to the question of democracy. The difference has to do with the fact that, while Tocqueville accepted democracy only reluctantly and showed interest in the idea of promoting a republican type of political culture and citizenship participation primarily as a way of minimizing damages that he believed democracy's tendency toward mediocrity and tyranny could inflict on individual liberties, and aristocracy's impulse to excellence; the theory of direct-deliberative *e*-democracy approaches the question of democracy from a decidedly optimistic and positive angle. Far from viewing democracy as an obstacle to human greatness and freedom, the theory finds in democracy a true path that would lead both the society and the human individual to excellence and liberty. Nurturing a republican type of political culture and facilitating citizenship participation are regarded by the theory as two essential components in promoting human excellence and liberty.

26. One should not lose sight of the fact that, for Rawls, the question of justice is not about democracy (or the political empowerment of the citizen directed toward her self-development), nor about the general welfare of society. Rather, like any other "standard liberal view," Rawls' interest in the question of justice is about the rights of individuals and maximizing their liberty or, as Quentin Skinner puts it, it is about demanding "a maximum degree of non-interference compatible with minimum demands of social life" (Skinner 1992, p.215).

27. Moreover, Macpherson's analysis of positive liberties also influenced the very formulation of the "conception of democracy as the political empowerment of the citizen." As was argued in Chapter 4, this conception corresponds to the first and third sense of positive liberties as categorized by Macpherson. On the question of the moral substance of democracy, as can be recalled from Chapter 12 of *The Betrayal of an Ideal*, Macpherson saw this substance primarily in substantive equality, and, through that, in collective sovereignty. The theory of direct-deliberative *e*-democracy, on the other hand, privileged the value inherent in the ideal of the citizens' direct participation in political decision-making as the main moral component of democracy.

28. Macpherson (1965), p.37.

29. Phrases in the quotation marks are borrowed from Joseph H. Carens in his commentary on Macpherson's "possessive individualism." Carens uses these phrases to characterize "possessive individualism" and not liberal democracy (Carens 1993, p.3).

30. The phrase in the quotation marks is borrowed from Barber (Barber 1984, p.20, original italics).

31. Claims of the moral neutrality of the state are deceptive; for the state's conduct, arguably, has moral force and implications, be they latent or overt, in intent or in consequence. This assertion of the theory of direct-deliberative *e*-democracy, however, does not obligate it to adopt a position that is often referred to as "perfectionism." The point of

this criticism is not that the state ought to be treated as an "ethical state" or as affirming some "*forms* of life" (i.e., some subjective or intersubjective norms of a specific ethical community) and rejecting others, but that the state should provide its citizens with the opportunity to choose their own forms or goals of life "freely." This in turn implies that the hegemonic position now enjoyed by the market-driven conceptions of life must give way to the hegemony of the conceptions of life in the civil society. (For a discussion of the question of "neutrality" vs. "perfectionism," see Mouffe (1993), pp.124-28. Mouffe uses Joseph Raz's contention that personal autonomy ought to be regarded as the basic value in the liberal-democratic state as an example of the perfectionist position.) However, the empirically-grounded charge that the claim of the moral neutrality of the state is deceptive or false does obligate the theory of direct-deliberative *e*-democracy to adopt the position that the state ought to be treated as a means for upholding a general conception of human development and for facilitating its citizens' pursuits of "morally valuable" ends; that the state ought to encourage citizens to aspire to universally acknowledged human-developmental ideals. (The ends or ideals in question are historically developed. That is to say, they are the ends that have withstood the test of history and have come to be recognized by the general consensus of past and present generations as ends desirable and valuable in themselves. In this sense, these ends are not only "universal," but also "intuitively" and "objectively" valid. Reverence and compassion for life, belief in and respect for the dignity and worth of each individual, individual autonomy, commitment to justice, amelioration of life's suffering, pacification and improvement of the conditions of human existence, and the positive development of the human individual (artistic, intellectual, moral, as well as physical) are among these ends.)

32. As can be recalled, the question of the domination of the civil society by the economic society and the latter's cultural hegemony was discussed in Chapter 4.

33. The phrase inside the quotation marks is borrowed from Dewey 1954[1927], p.184. Here Dewey has in mind the despotism of "the machine age" (i.e., the economy of the "Great Society" of industrial capitalism).

34. Marx's words in praising the democracy among the Greeks, in a letter written to Arnold Ruge in May 1843 (Padover 1979, p.25). See Chapter 4 of *The Betrayal of an Ideal* for further details.

35. Dewey (1954[1927]), p.184.

Selected Bibliography

Addams, Jane. *Democracy and Social Ethics*. Cambridge, MA: The Belknap Press of Harvard University Press, 1964 [1902].

Albert, Michel. *Capitalism vs. Capitalism: How America's Obsession with Individual Achievement and Short-term Profit Has Led It to the Brink of Collapse*. New York: Four Wall Eight Windows, 1993.

Alexander, Cynthia J., and Leslie A. Pal, eds. *Digital Democracy: Policy and Politics in the Wired World*. Oxford: Oxford University Press, 1998.

Allen, Anita L., and Milton C. Regan, Jr., eds. *Debating Democracy's Discontent: Essay on American politics, Law, and Public Philosophy*. Oxford: Oxford University Press, 1998.

Amy, Douglas J. *Behind the Ballot Box: A Citizen's Guide to Voting Systems*. Westport, CT: Praeger, 2000.

Ankersmit, F. R. *Political Representation*. Stanford: Stanford University Press, 2002.

Appleby, Joyce, Lynn Hunt, and Margaret Jacobs. *Telling the Truth about History*. New York: W. W. Norton and Company, 1994.

Archer, Robin. *Economic Democracy: The Politics of Feasible Socialism*. Oxford: Clarendon Press, 1995.

Archer, Robin, Diemut Bubeck, Hanjo Glock, Lesley Jacobs, Seth Moglen, Adam Steinhouse, and Daniel Weinstock, eds. *Out of Apathy: Voices of the New Left Thirty Years On*. London: Verso, 1989.

Aristotle. *The Politics of Aristotle*. Translated and with an introduction, notes and appendixes by Ernest Barker. Oxford: Oxford University Press, 1958.

Arrow, Kenneth. *Social Choice and Individual Values*. New Haven: Yale University Press, 1963.

Arrow, Kenneth, Amartya Sen, and Kotaro Suzumura, eds. *Social Choice Re-Examined: Proceedings of the IEA Conference Held at Schloss Hernstein, Berndorf, Vienna, Austria*, volumes 1 and 2. London: Macmillan Press LTD, 1996.

Arthur, John, ed. *Democracy: Theory and Practice*. Belmont, CA: Wadsworth Publishing Company, 1992.

Avon, Dan, and Anver De-Shalit, eds. *Liberalism and Its Practice*. London: Routledge, 1999.
Bachrach, Peter, and Aryeh Botwinick, eds. *Power and Empowerment: A Radical Theory of Participatory Democracy*. Philadelphia: Temple University Press, 1992.
Baker, C. Edwin. *Media, Markets, and Democracy*. Cambridge: Cambridge University Press, 2002.
Baker, Earnest. *The Political Thought of Plato and Aristotle*. New York: Dover Publications, Inc., 1959.
Baran, Paul, and Paul Sweezy. *Monopoly Capital: An Essay on the American Economic and Social Order*. New York: Monthly Press Review, 1968.
Barber, Benjamin R. *Strong Democracy*. Berkeley: University of California Press, 1984.
——. *A Passion for Democracy*. Princeton: Princeton University Press, 1998.
——. "Three Scenarios for the Future of Technology and Strong Democracy." *Political Science Quarterly*, vol.113, no.4 (1998-99): 573-89.
Beard, Charles. *An Economic Interpretation of the Constitution of the United States*. New York: Macmillan Company, 1935.
Beaud, Michel. *A History of Capitalism 1500-2000*. Translated and edited by Tom Dickman and Anny Lefebvre. New York: Monthly Review Press, 2001.
Becker, Barbara, and Josef Wehner. "Electronic Networks and Civil Society: Reflections on Structural Changes in the Public Sphere." In *Culture, Technology, Communication: Toward an Intercultural Global Village*, edited by Charles Ess and Fay Sudweeks. Albany, NY: State University of New York Press, 2001.
Behrouzi, Majid. *Democracy as the Political Empowerment of the People: The Betrayal of an Ideal*. Lanham, MD: Lexington Books, 2005.
Beitzinger, A. J. *A History of American Political Thought*. New York: Dodd, Mead and Company, 1972.
Bender, Frederic L. *The Betrayal of Marx*. New York: Harper & Row Publishers, 1975.
Benello, C. George. *From the Ground Up: Essays on Grassroots and Workplace Democracy*. Boston: South End Press, 1992.
Berlin, Isaiah. *Four Essays on Liberty*. Oxford: Oxford University Press, 1969.
Berry, Christopher J. *The Idea of a Democratic Community*. New York: St. Martin's Press, 1989.
Birch, Anthony. H. *Representation*. New York: Praeger, 1971.
Black, Duncan, and R.A. Newing. *The Theory of Committees and Elections* and *Committee Decisions With Complementary Valuation*. Revised second ed. Boston: Kluwer Academic Publishers, 1998.
Blaug, Richardo. "New Theories of Discursive Democracy: A User's Guide." *Philosophy and Social Criticism*, vol.22, no.1 (1996): 49-80.
——. *Democracy, Real and Ideal: Discourse Ethics and Radical Politics*. Albany, NY: State University of New York Press, 1999.

Boardman, John, Jasper Griffin, and Oswyn Murray, eds. *The Oxford History of the Classical World*. Oxford: Oxford University Press, 1986.

Bloom, Allan. "Rousseau's Critique of Liberal Constitutionalism." In *The Legacy of Rousseau*, edited by Clifford Orwin and Nathan Tarcov. Chicago: The University of Chicago Press, 1997.

Bohman, James. "Survey Article: The Coming of Age of Deliberative Democracy." *The Journal of Political Philosophy,* vol.6, no.4 (1998): 400-425.

Bonner, Robert J. *Aspects of Athenian Democracy*. New York: Russell & Russell, 1967.

Borgmann, Albert. *Technology and the Character of the Contemporary Life*. Chicago: The University of Chicago Press, 1984.

——. *Crossing the Postmodern Divide*. Chicago: The University of Chicago Press, 1992.

——. *Holding on to Reality: The Nature of Information at the Turn of the Millennium*. Chicago: The University of Chicago Press, 1999.

Bottomore, Tom, Laurence Harris, V. G. Kiernan, and Ralph Miliband, eds. *A Dictionary of Marxist Thought*. Cambridge, MA: Harvard University Press, 1983.

Bowles, Samuel, and Herbert Gintis. *Democracy and Capitalism: Property, Community, and the Contradictions of Modern Social Thought*. New York: Basic Books, 1986.

Brams, Steven J., and Peter C. Fishbum. *Approval Voting*. Boston: Birkhauser, 1983.

Brink, Bert Van Den. *The Tragedy of Liberalism: An Alternative Defense of a Political Tradition*. Albany: State University of New York Press, 2000.

Broder, David S. *Democracy Derailed: Initiative Campaigns and the Power of Money*. New York: Harcourt Inc., 2000.

Brown, A. "Reconstructing the Soviet Political System." In *Chronicle of a Revolution: A Western- Soviet Inquiry Into Perestroika*, edited by A. Brumberg. New York: Pantheon, 1990.

Browning, Graeme. *Electronic Democracy: Using the Internet to Transfer American Politics*. Medford, NJ: Cyber Age Books, Information Today Incorporated, 2002.

Buchanan, James. "Social Choice, Democracy, and Free Markets." *Journal of Political Economy*, April 1954, 62(2), pp.114-23.

Buchanan, James M., and Gordon Tullock. *The Calculus of Consent: Logical Foundations of Constitutional Democracy*. Ann Arbor: University of Michigan Press, 1965.

Budge, Ian. *The New Challenge of Direct Democracy*. Cambridge: Polity Press, 1996.

Burke, Edmund. *The Writing and Speeches of the Right Honorable Edmund Burke*. Beaconsfield edition in twelve volumes. Boston: Little, Brown and Company, 1901.

Burnheim, John. *Is Democracy Possible?* Berkeley: University of California Press, 1985.

Calhoun, Craig. "The Public Good as a Social and Cultural Project." In *Private Action and the Public Good*, edited by Walter W. Powell and Elizabeth S. Clemens. New Haven: Yale University Press, 1998.

Carens, Joseph H., ed. *Democracy and Possessive Individualism: The Intellectual Legacy of C. B. Macpherson*. Albany, NY: State University of New York Press, 1993.

Chapman, John W., and I. Shapiro, eds. *Democratic Community*. New York: New York University Press, 1993.

Chomsky, Noam. *Profit Over People*. New York: Seven Stories Press, 1999.

Cladis, Mark S. *Public Visions, Private Lives: Rousseau, Religion, and 21st-Century Democracy*. Oxford: Oxford University Press, 2003.

Cohen, Carl, ed. *Communism, Fascism, and Democracy: The Theoretical Foundations*. New York: McGraw-Hill, 1997.

Cohen, Joshua. "Democracy and Liberty." In *Deliberative Democracy*, edited by John Elster. Cambridge: Cambridge University Press, 1998.

———. "An Epistemic Conception of Democracy." *Ethics*, vol.97 (October 1986): 26-38.

———. "Deliberation and Democratic Legitimacy." In *Deliberative Democracy: Essays on Reason and Politics*, edited by James Bohman and William Rehg. Cambridge, MA: MIT Press, 1997a.

———. "Procedure and Substance in Deliberative Democracy." In *Deliberative Democracy: Essays on Reason and Politics*, edited by James Bohman and William Rehg. Cambridge, MA: MIT Press, 1997b.

Cohen, Joshua, and Joel Rogers. "Associative Democracy." In *Market Socialism: The Current Debate*, edited by Pranab K. Bardhan and John E. Roemer. Oxford: Oxford University Press, 1993.

———. *Associations and Democracy: The Utopian Project*, vol.1, edited by Erik Olin Wright. London: Verso, 1995.

———. "My Utopia or Yours?" In *Equal Shares: Making Market Socialism Work the Real Utopias Project*, vol.II, by John E. Roemer, edited and introduced by Erik Olin Wright. London: Verso, 1996.

Coleman Jules L., and John Ferejohn. "Democracy and Social Choice." *Ethics*, vol.97 (October 1986): 6-25.

Colletti, L. *From Rousseau to Lenin*. London: New Left Books, 1972.

Constant, Benjamin. *The Political Writings of Benjamin Constant*. Edited by Biancamaria Fontona. Cambridge and New York: Cambridge University Press, 1988.

Copp, David, Jean Hampton, and John Roemer, eds. *The Ideal of Democracy*. Cambridge: Cambridge University Press, 1993.

Corlett, J. Angelo. "Marx and Rights." *Dialogue* (Canada), XXXIII (1994).

Cowling, Mark, and Paul Reynolds, eds. *Marxism, the Millennium, and Beyond*. London: Palgrave Publishers Ltd., 2000.

Craig, Edward, gen. ed. *Routledge Encyclopedia of Philosophy*. London: Routledge, 1998.

Craig, Stephen C., ed. *Broken Contract? Changing Relations Between Americans and Their Government*. Boulder, CO: Westview Press, 1996.

Cunningham, Frank. *Democratic Theory and Socialism*. Cambridge: Cambridge University Press, 1987.

———. *The Real World of Democracy Revisited*. Atlantic Highlands, NJ: Humanities Press, 1994.

———. *Theories of Democracy: A Critical Introduction*. London: Routledge, 2002.

Dagger, Richard. *Civic Virtues: Rights, Citizenship, and Republican Liberalism*. New York: Oxford University Press, 1997.

Dahl, Robert A. *A Preface to Economic Theory*. Chicago: The University of Chicago Press, 1956.

———. *Democracy and Its Critics*. New Haven, CT: Yale University Press, 1989.

———. *On Democracy*. New Haven, CT: Yale University Press, 1998.

Dahl, Robert, Ian Shapiro, Jose Antonio Cheibub. *The Democracy Source Book*. Cambridge, MA: MIT Press, 2003a.

Dalton, Russell J., Wilhelm Burklin, and Andrew Drummond. "Public Opinion and Direct Democracy." *Journal of Democracy*, vol.12, no.4 (October 2001).

Dante, Germino. *Machiavelli to Marx: Modern Western Political Thought*. Chicago: The University of Chicago Press, 1972.

Davis, Richard. *The Web of Politics: The Internet's Impact on the American Political System*. Oxford: Oxford University Press, 1999.

Delanty, Gerard. *Community*. London: Routledge, 2003.

DeLong, Howard. *A Refutation of Arrow's Theorem*. New York: University Press of America, 1991.

DeMarco, Joseph P., and Samuel A. Richmond. "The Mutuality of Liberty, Equality, and Fraternity." *Journal of Social Philosophy* 17 (Fall 1986): 7-12.

Denning, S. Lance. *The Practice of Workplace Participation*. Westport, CT: Quorum Books, 1998.

Deutscher, Isaac. *The Profit Unarmed, Trotsky: 1921-1929*. London: Oxford University Press, 1959.

Dew, John R. *Empowerment and Democracy in the Workplace*. Westport, CT: Quorum Books, 1997.

Dewey, John. *Philosophy of Education*. Littlefield: Adams and Company, 1975.

———. *Liberalism and Social Action*. New York: Perigee Books, 1980.

———. *The Public and Its Problems*. Chicago: Swallow Press, 1954 [1927].

———. "The Ethics of Democracy." In *Jon Dewey: The Political Writings*, edited and introduced by Debra Morris and Ian Shapiro. Indianapolis, IN: Hackett Publishing Company, 1993 [1888].

Donnelly, David, Janice Fine, and Ellen S. Miller. *Are Elections for Sale?* Boston: Beacon Press, 2001.

Downs, Anthony. *An Economic Theory of Democracy*. New York: Harper and Row, 1957.

Draper, Hal. *The "Dictatorship of the Proletariat" From Marx to Lenin*. New York: Monthly Review Press, 1987.
Drew, Elizabeth. *The Corruption of American Politics: What Went Wrong and Why*. New York: The Overlook Press, 2000.
——. *Politics and Money*. New York: Macmillan Publishing Company, 1983.
Dryzek, John S. *Deliberative Democracy and Beyond: Liberals, Critics, Contestations*. Oxford: Oxford University Press, 2000.
——. "Legitimacy and Economy in Deliberative Democracy." *Political Theory*, vol.29, no.5 (October 2001): 651-69.
DuBoff, Richard B. *Accumulation and Power: An Economic History of the United States*. Armonk, NY: M. E. Sharpe Inc., 1989.
Dugger, Ronnie. "How They Could Steal the Election This Time." *The Nation* (August 16-23, 2004): 11-24.
Dummett, Michael. *Voting Procedures*. Oxford: Clarendon Press, 1984.
Duncan, Christopher M. *The Anti-Federalists and Early American Political Thought*. DeKalb, IL: Northern Illinois University Press, 1995.
Dunn, John, ed. *Democracy the Unfinished Journey: 508 BC to AD 1993*. Oxford: Oxford University Press, 1992.
Dworkin, Ronald. *A Matter of Principle*. Cambridge, MA: Harvard University Press, 1985.
——. *Sovereign Virtue*. Cambridge, MA: Harvard University Press, 2000.
Dye, Thomas R. *Who's Running America? The Clinton Years*. Sixth edition. Englewood Cliffs: Prentice Hall, 1995.
Edwards, Paul (chief editor). *The Encyclopedia of Philosophy*. New York: Macmillan Publishing Company, 1967.
Eisenberg, Avigail I. *Reconstructing Political Liberalism*. Albany: State University of New York Press, 1995.
Ellul, Jacques. *The Technological Society* (1954). Translated by John Wilkinson. New York: Vintage Books, 1964 (1954).
——. "Technology and Democracy." In *Democracy in a Technological Society*, edited by Langdon Winner. Boston: Kluwer Academic Publishers, 1992.
Elster, Jon, and Aanurd Hylland, eds. *Foundations of Social Choice Theory*. Cambridge: Cambridge University Press, 1986.
Elster, Jon, ed. *Deliberative Democracy*. Cambridge: Cambridge University Press, 1998.
——. "Possibility of Rational Politics." In *Political Theory Today*, edited by David Held. Stanford: Stanford University Press, 1991.
Estlund, David M. "Who's Afraid of Deliberative Democracy? On the Strategic/Deliberative Dichotomy in Recent Constitutional Jurisprudence." *Texas Law Review*, vol.71 (1993): 1437-77.
Etzioni, Amitai. *New Communitarian Thinking: Persons, Virtues, Institutions, and Communities*. Charlottesville: University Press of Virginia, 1995.
Farber, Samuel. *Before Stalinism: The Rise and Fall of Soviet Democracy*. London: Verso, 1990.

Feenberg, Andrew. *Critical Theory of Technology*. Oxford: Oxford University Press, 1991.
——. *Alternative Modernity*. Berkeley: University of California Press, 1995.
——. *Questioning Technology*. London: Routledge, 1999.
——. *Transforming Technology: A Critical Theory Revised*. Oxford: Oxford University Press, 2002.
Feenberg, Andrew, and Alastair Hannay, eds. *Technology and Politics of Knowledge*. Bloomington, IN: Indiana University Press, 1995.
Femia, Joseph V. *Marxism and Democracy*. Oxford: Clarendon Press, 1993.
——. "Complexity and Deliberative Democracy." *Inquiry* 39 (1996): 359-97.
Fishkin, James S. *Democracy and Deliberation*. New Haven and London: Yale University Press, 1991.
Flower, Robert B. *The Dance with Community*. Lawrence: University Press of Kansas, 1991.
Franklin, Ursula. *The Real World of Technology*. Toronto: CBC Enterprises, 1990.
Freeman, Samuel. "Deliberative Democracy: A Sympathetic Comment." *Philosophy and Public Affairs* 29, no.4 (Fall 2000): 371-418.
Friedland, L. A. "Electronic Democracy and the New Citizenship." *Media Culture & Society* 18:2 (April 1996): 185-212.
Friedland, Edward I., and J. Cimbala. "Process and Paradox: The Significance of Arrow's Theorem." *Theory and Decision* 4 (1973): 51-64.
Gendin, Sidney. "Why Arrow's Impossibility Theorem Is Invalid." *Journal of Social Philosophy*, vol.25, no.1 (Spring 1994): 144-59.
Germino, Dante. *Machiavelli to Marx: Modern Western Political Thought*. Chicago: The University of Chicago Press, 1979.
Golding, Sue. *Gramsci's Democratic Theory: Contributions to a Post-Liberal Democracy*. Toronto: University of Toronto Press, 1992.
Gould, Carol C. *Rethinking Democracy*. Cambridge: Cambridge University Press, 1988.
Grady, Robert C. *Restoring Real Representation*. Urbana: University of Illinois Press, 1993.
Gramsci, A. *Selections from Prison Notebooks*. New York: International Publishers, 1971.
Gray, John. *Liberalism*. Minneapolis: University of Minnesota Press, 1995.
Green, Judith M. *Deep Democracy: Community, Diversity, and Transformation*. Lanham, MD: Rowman & Littlefield Publishers, Inc., 1999.
Green, Leslie. "Dictators and Democracies." *Analysis*, vol.43, no.1 (January 1983): 58-59.
Green, Mark. *Selling Out: How Big Corporate Money Buys Elections, Rams Through Legislation, and Betrays Our Democracy*. New York: Regan Books, 2004.
Greenberg, Edward S. *Workplace Democracy: The Political Effects of Participation*. Ithaca: Cornell University Press, 1986.

Greenberg, Stanley B. *The Two Americas: Our Current Political Deadlock and How to Break it.* New York: Thomas Dunne Books, 2004.
Grossman, Lawrence, K. *The Electronic Republic: Reshaping Democracy in the Information Age.* New York: Viking Press, 1995.
Gundersen, Adolf G. *The Socratic Citizens: A Theory of Deliberative Democracy.* Lanham, MD: Lexington Books, 2000.
Gutmann, Amy, and Dennis Thompson. *Democracy and Disagreement.* Cambridge, MA: Harvard University Press, 1996.
Habermas, Jurgen. *Legitimation Crisis.* Translated by Thomas McCarthy. Boston: Beacon Press, 1975.
——. "Three Normative Models of Democracy." *Constellations,* vol.1, no.1 (1994): 1-10.
——. "Human Rights and Popular Sovereignty: The Liberal and Republican Versions." In *Human Rights Law,* edited by Philip Alston. New York: New York University Press, 1996.
——. "Popular Sovereignty and Procedure." In *Deliberative Democracy: Essays on Reason and Politics,* edited by James Bohman and William Rehg. Cambridge, MA: MIT Press, 1997.
——. *Between Facts and Norms.* Cambridge, MA: MIT Press, 1998.
Hamilton, Alexander, James Madison, and John Jay. *The Federalist.* Edited by Benjamin Fletcher Wright. Cambridge, MA: The Belknap Press of Harvard University Press, 1961.
Haumptmann, Emily. *Putting Choice Before Democracy.* Albany: State University of New York Press, 1996.
Held, David. *Models of Democracy.* Palo Alto, CA: Stanford University Press, 1989.
Hertz, Noreena. *The Silent Takeover: Global Capitalism and the Death of Democracy.* London: Heinemann, 2001.
——. *Prospects for Democracy: North South East West.* Stanford: Stanford University Press, 1993.
——. *Democracy and the Global Order.* Cambridge: Polity Press, 1995.
Hibbing, John, R., and Elizabeth Theiss-Morse. *What Is It About Government That Americans Dislike?* London: Cambridge University Press, 2001.
Hightower, Jim. *Thieves in High Places: They've Stolen Our Country and It's Time to Take It Back.* New York: Viking, 2003.
Hirst, Paul. *Representative Democracy and Its Limits.* Cambridge: Polity Press, 1990.
Hobbes, Thomas. *Leviathan* (1651). Edited by Richard Tuck. Cambridge: Cambridge University Press, 1996.
Hoffman, Ronald, and Peter J. Albert, eds. In *The Transforming Hand of Revolution: Reconsidering the American Revolution as a Social Movement.* Charlottesville: The University Press of Virginia, 1996.
Howard, Michael W. *Self-Management and the Crisis of Socialism: The Rose in the Fist of the Present.* Lanham, MD: Rowman and Littlefield Publishers, Inc., 2000.

——, ed. *Socialism*. Amherst, MA: Humanity Books, 2001.

Howard, Rhoda, E. *Human Rights and Search for Community*. Oxford: Westview Press, 1995.

Huffington, Arianna. *How to Overthrow the Government*. New York: Harper Collins Publishers, 2000.

——. *Pigs at the Trough: How Corporate Greed and Political Corruption Are Undermining America*. New York: Crown Publishers, 2003.

Ihde, Don. *Technology and the Lifeworld: From Garden to Earth*. Bloomington, IN: Indiana University Press, 1990.

Isaac, Jeffrey C. *Democracy in Dark Times*. Ithaca, NY: Cornell University Press, 1988.

Jacobs, Lesley A. *An Introduction to Modern Political Philosophy*. Upper Saddle River, NJ: Prentice Hall, 1997.

——. "Market Socialism and Non-Utopian Marxist Theory." *Philosophy of Social Sciences,* vol.29, no.4 (December 1999): 527-39.

Jaggar, Allison M. "Multicultural Democracy." *The Journal of Political Philosophy,* vol.7, no.3 (1999): 308-29.

Johnson, D. G. *Computer Ethics*. Englewood Cliffs, NJ: Prentice Hall, 1994.

Jones, A. H. M. *Athenian Democracy*. Oxford: Basil Blackwell, 1964.

Judis, John B. *The Paradox of American Democracy*. New York: Routledge, 2000.

Kamarck, Elaine, and Joseph Nye, eds. *Governance.Com Democracy in the Information Age*. Washington, D.C.: Brooking Institution Press, 2002.

Kant, Immanuel. *Foundations of the Metaphysics of Morals* (1785). Translated and Introduced by Lewis White Beck. New York: Macmillan Publishing Company, 1959.

Kautz, Steven. *Liberalism and Community*. Ithaca: Cornell University Press, 1995.

Kegley, Jacquelyn A. K. *Genuine Individuals and Genuine Communities*. Nashville: Vanderbilt University Press, 1997.

Kenez, Peter. *The Birth of the Propaganda State*. Cambridge: Cambridge University Press, 1985.

Knei-Paz, Baruch. "Was George Orwell Wrong?" *Dissent* 42 (2) (Spring 1995): 266-69.

Knight, Jack, and James Johnson. "Aggregation and Deliberation: On the Possibility of Democratic Legitimacy." *Political Theory,* vol.22, no.2 (May 1994): 277-96.

Kochler, Hans. *The Crisis of Representative Democracy*. Frankfurt: Verlag Peter Lang, 1987.

Krimerman, Len, and Frank Lindenfeld, eds. *When Workers Decide: Workplace Democracy Takes Root in North America*. Philadelphia: New Society Publishers, 1992.

Kulikoff, Allan. *The Agrarian Origins of American Capitalism*. Charlottesville: University Press of Virginia, 1992.

Kymlicka, Will. *Liberalism, Community, and Culture*. Oxford: Clarendon Press, 1981.

——. *Contemporary Political Philosophy*. New York: Oxford University Press, 1990.

——. *Finding Our Way: Rethinking Ethnocultural Relations in Canada*. Oxford: Oxford University Press, 1998.

——. "Liberal Egalitarianism and Civic Republicanism: Friends or Enemies?" In *Debating Democracy's Discontent: Essay on American Politics, Law, and Public Philosophy*, edited by Anita L. Allen and Milton C. Regan. Oxford: Oxford University Press, 1998.

——. "American Multiculturalism and the 'Nation Within.'" In *Political Theory and the Rights of Indigenous Peoples*, edited by Ivison Duncan, Paul Patton, and Will Sanders. Cambridge: Cambridge University Press, 2000.

——. *Contemporary Political Philosophy*. Oxford: Oxford University Press, 2002.

Lane, David. *Leninism: A Sociological Interpretation*. Cambridge: Cambridge University Press, 1981.

Lenin, V. I. *Selected Works in Two Volumes*. Moscow: Foreign Languages Publishing House, 1952.

——. *Collected Works*. Moscow: Progress Publishers, 1976.

Levine, Andrew. *The Politics of Autonomy: A Kantian Reading of Rousseau's Social Contract*. Amherst: University of Massachusetts Press, 1976.

——. *Liberal Democracy: A Critique of Its Theory*. New York: Columbia University Press, 1981.

——. *Arguing for Socialism*. Boston, London: Routledge & Kegan Paul, 1984.

——. *The General Will: Rousseau, Marx, Communism*. Cambridge: Cambridge University Press, 1993.

——. "Democratic Corporatism and/versus Socialism." In *Associations and Democracy: The Utopian Project*, vol.1. Edited by Joshua Cohen and Joel Rogers. London: Verso, 1995.

——. *Engaging Political Philosophy: From Hobbes to Rawls*. Oxford: Blackwell Publishers, 2002.

Lichterman, Paul. *The Search for Community: American Activists Reinventing Commitment*. Cambridge: Cambridge University Press, 1996.

Lindsay, Peter. *Creative Individualism*. Albany: State University of New York Press, 1996.

Lipset, Seymour Martin (chief editor). *The Encyclopedia of Democracy*. In 4 vols. Washington, D.C.: Congressional Quarterly Inc., 1995.

List, Christian, and Robert E. Goodin. "Epistemic Democracy: Generalizing the Condorcet Jury Theorem." *The Journal of Political Philosophy*, vol.9, no.3 (2001): 277-306.

Lopston, Peter. *Theories of Human Nature*. Peterborough, ON: Broadview Press, 1995.

Lovell, David W. *From Marx to Lenin: An Evaluation of Marx's Responsibility for Soviet Authoritarianism*. Cambridge: Cambridge University Press, 1984.

Lummis, C. Douglas. *Radical Democracy*. Ithaca, NY: Cornell University Press, 1996.

Luxemburg, Rosa. *Rosa Luxemburg Speaks*. Edited by Mary-Alice Waters. New York: Pathfinder Press, Inc., 1970.

Macedo, Stephen, ed. *Deliberative Politics: Essays on Democracy and Disagreement*. Oxford: Oxford University Press, 1999.

Machiavelli, Niccolo. *The Prince*. Edited with an introduction by Peter Bondanella, translated by Peter Bondanella and Mark Musa. Oxford: Oxford University Press, 1984.

MacKay, Alfred, F. *Arrow's Theorem, The Paradox of Social Choice: A Case Study in the Philosophy of Economics*. New Haven: Yale University Press, 1980.

Mackie, Gerry. "Saving Democracy from Political Science." In *The Democracy Source Book*, edited by Robert Dahl, Ian Shapiro, and Jose Antonio Cheibub. Cambridge, MA: MIT Press, 2003a.

——. *Democracy Defended*. Cambridge: Cambridge University Press, 2003b.

Macpherson, C. B. *The Real World of Democracy*. Toronto: House of Anansi Press, 1987 [1965].

——. *Democratic Theory: Essays in Retrieval*. Oxford: Clarendon Press, 1973.

——. *The Life and Times of Liberal Democracy*. Oxford: Oxford University Press, 1977.

Manin, Bernard. *The Principles of Representative Government*. Cambridge: Cambridge University Press, 1997.

Mansbridge, Jane. *Beyond Adversary Democracy*. New York: Basic Books, 1980.

——. "Reconstructing Democracy." In *Revisioning the Political: Feminist Reconstructions of Traditional Concepts in Western Political Theory*, edited by Nancy Hirschmann and D. Stefano. Boulder, CO: Westview Press, 1996.

——. "On the Contested Nature of the Public Good." In *Private Action and the Public Good*, edited by Walter W. Powell and Elisabeth S. Clemens. New Haven: Yale University Press, 1998.

Marcuse, Herbert. *Reason and Revolution: Hegel and the Rise of Social Theory* (1954). Second ed. with supplementary chapter. Atlantic Highlands, NJ: Humanities Press, Inc., 1983.

——. *Eros and Civilization: A Philosophical Inquiry Into Freud* (1955). Boston: Beacon Press, 1966.

——. *One Dimensional Man: Studies in the Ideology of Advanced Industrial Society*. Boston: Beacon Press, 1964.

Marien, Michael. "New Communications Technologies: A Survey of Impacts and Issues." *Telecommunications Policy*, vol.20, no.5 (1996): 375-87.

Marx, Karl. *Karl Marx: Early Writings*. Translated and edited by T. B. Bottomore. New York: McGraw Hill, 1963.

——. *Critique of Hegel's "Philosophy of Right."* Edited by Joseph O'Malley. Cambridge: Cambridge University Press, 1970 [1859].

——. *The Letters of Karl Marx*. Edited by Saul K. Padover. Englewood Cliffs, NJ: Prentice Hall, 1979.
Marx, Karl, and Fredrick Engels. *Selected Works in Three Volumes*. Moscow: Progress Publishers, 1977.
——. *The German Ideology*. Moscow: Progress Publishers, 1976.
——. *Collected Works*. New York: International Publishers, 1975.
McChesney, Robert W., Ellen Meiksins Wood, and John Bellamy Foster, eds. *Capitalism and the Information Age: The Political Economy of the Global Communication Revolution*. New York: Monthly Review Press, 1998.
McClure, Kristie. "On the Subject of Rights: Pluralism, Plurality and Political Identity." In *Dimensions of Radical Democracy: Pluralism, Citizenship, Community*. Eedited by Chantal Mouffe. London: Verso, 1992.
McCullagh, Karen. "E-democracy: Potential for Political Revolution?" *International Journal of Law and Information Technology*, vol.11, no.2 (2003): 149-61.
McIver, William J. Jr., and Ahmed K. Elmagarmid, eds. *Advances in Digital Government: Technology, Human Factors, and Policy*. Boston: Kluwer Academic Publishers, 2002.
McLean, Iain. *Dealing in Votes: Interactions Between Politicians and Voters in Britain and the USA*. Oxford: Martin Robertson, 1982.
——. *Public Choice: An Introduction*. Oxford: Basil Blackwell, 1987.
——. *Democracy and New Technology*. Cambridge: Polity Press, 1989.
——. "Rational Choice and Politics." *Political Studies*, vol.xxxix (1991a): 496-512.
——. "Forms of Representation and Systems of Voting." In *Political Theory Today*, edited by David Held. Stanford: Stanford University Press,1991b.
McLean, Iain, and Arnold B. Urken., eds. *Classics of Social Choice*. Ann Arbor: University of Michigan Press, 1995.
Melman, Seymour. *After Capitalism: From Managerialism to Workplace Democracy*. New York: Alfred A. Knopf, 2001.
Melzer, Arthur M. *The Natural Goodness of Man: On the System of Rousseau's Thought*. Chicago: The University of Chicago Press, 1990.
Miliband, Ralph. *The State in Capitalist Society*. London: Quartet Books, 1980.
Mill, John Stuart. *Considerations on Representative Government* (1861). Edited by Currin V. Shields. New York: The Bobbs-Merrill Company, Inc., 1958.
Miller, David. *Market, State, and Community: Theoretical Foundations of Market Socialism*. Oxford: Clarendon Press, 1989.
——. *Citizenship and National Identity*. Cambridge: Polity Press, 2000.
Miller, James. "*Democracy Is in the Streets*": *From Port Huron to the Siege of Chicago*. New York: Simon and Schuster, 1987.
Miller, Joshua. *The Rise and Fall of Democracy in Early America, 1630-1789*. University Park: The Pennsylvania State University Press, 1991.
Mitcham, Carl. *Thinking Through Technology: The Path Between Engineering and Technology*. Chicago: The University of Chicago Press, 1994.

Montesquieu, Charles-Louis de Secondat. *The Spirit of the Laws*. Translated and edited by Anne M. Cohler, Basia Carolyn Miller, and Harold Samuel Stone. Cambridge: Cambridge University Press, 1989.

Morera, Esteve. *Gramsci's Historicism*. London: Routledge, 1990.

——. "Gramsci and Democracy." *Canadian Journal of Political Science*, XXIII:1 (March 1990): 23-37.

Morris, Dick. *Vote.com*. Los Angeles: Renaissance Books, 1999.

Morrison, John. "The Case Against Constitutional Reform?" *Journal of Law and Society*, vol.25, no.4 (December 1998): 510-35.

Mossberger, Karen, Caroline J. Tolbert, and Mary Stansbury. *Virtual Inequality: Beyond the Digital Divide*. Washington D.C.: Georgetown University Press, 2003.

Mouffe, Chantal, and Ernesto Laclua. *Hegemony and Socialist Strategy: Toward a Radical Democratic Politics*. London: Verso, 1985.

Mouffe, Chantal, ed. *Dimensions of Radical Democracy: Pluralism, Citizenship, Community*. London: Verso, 1992.

——. *The Return of the Political*. London: Verso, 1993.

Mueller, Dennis. *Public Choice II*. Cambridge: Cambridge University Press, 1989.

Mukherjee, Rabin. *Democracy: A Failure, Shefocracy: The Solution for Human Welfare*. New York: University Press of America, 2000.

Murakami, Yasuske. *Logic and Social Choice*. London: Routledge & Kegan Paul Ltd., 1968.

Nader, Ralph. *Crashing the Party: Taking on the Corporate Government in an Age of Surrender*. New York: St. Martin's Press, 2002.

Nedelsky, Jennifer. *Private Property and the Limits of American Constitutionalism*. Chicago: The University of Chicago Press, 1990.

Neidleman, Jason A. *The General Will Is Citizenship: Inquiries into French Political Thought*. New York: Rowman and Littlefield Publishing Co., 2001.

Newman, Bruce I. *The Mass Marketing of Politics: Democracy in an Age of Manufactured Images*. Thousand Oaks, CA: Sage Publications, 1999.

Nino, Carlos Santiago. *The Constitution of Deliberative Democracy*. New Haven: New York University Press, 1996.

Norris, Pippa. *Digital Divide: Civic Engagement, Information Poverty, and the Internet Worldwide*. Cambridge: Cambridge University Press, 2001.

Nun, Jose. *Democracy: Government of the People or Government of the Politicians?* Lanham, MD: Rowman & Littlefield Publishers, 2003.

Nurmi, Hannu. *Comparing Voting Systems*. Dordrecht: D. Reidel Publishing Company, 1987.

——. *Voting Paradoxes and How to Deal with Them*. New York: Springer, 1999.

Ogden, M. R. "Politics in a Parallel Universe: Is There a Future for Cyberdemocracy?" *Futures*, vol.26, no.7 (1994): 713-29.

——. "Electronic Power to People: Who Is Technology's Keeper on the Cyberspace Frontier?" *Technological Forecasting & Social Change* 52 (2-3) (June-July 1996): 119-33.

Ollman, Bertell, ed. *Market Socialism: The Debate Among the Socialists*. New York: Routledge,1998.

O'Shaughnessy, Nicholas, J. *The Phenomenon of Political Marketing*. London: Macmillan Press, 1990.

Pagano, Ugo, and Robert Rowthorn, eds. *Democracy and Efficiency in the Economic Enterprise*. London: Routledge, 1996.

Palast, Greg. *The Best Democracy Money Can Buy*. London: Pluto Press, 2002.

Pangle, Thomas L. *Montesquieu's Philosophy of Liberalism: A Commentary on the Spirit of the Laws*. Chicago: The University of Chicago Press, 1973

Parenti, Michael. *Democracy for the Few*. New York: St. Martin's Press, 1996.

Pateman, Carole. *Participation and Democratic Theory*. Cambridge: Cambridge University Press, 1970.

——. "Social Choice or Democracy? A Comment on Coleman and Ferejohn." *Ethics*, vol.97 (October 1986): 39-46.

Peffer, R. G. *Marxism, Morality, and Social Justice*. Princeton: Princeton University Press, 1990.

Pettit, Philip. "Reworking Sandel's Republicanism." In *Debating Democracy's Discontent: Essays on American Politics, Law, and Public Philosophy*, edited by Anita L. Allen and Milton C. Regan. Oxford: Oxford University Press, 1998.

——. "The Virtual Reality of *Homo Economicus*." In *The Economic World View*, edited by Uskali Maki. Cambridge: Cambridge University Press, 2001.

——. "Keeping Republican Freedom Simple: On a Difference with Quentin Skinner." *Political Theory*, vol.30, no.3 (June 2002): 339-56.

——. *Republicanism: A Theory of Freedom and Government*. Oxford: Oxford University Press, 1997.

Phillips, Kevin. *Wealth and Democracy: A Political History of the American Rich*. New York: Broadway Books, 2002.

——. *American Dynasty*. New York: Viking, 2004.

Pitkin, Hanna Fenichel. *The Concept of Representation*. Berkeley: University of California Press, 1967.

Plato. *The Republic of Plato*. Edited by Francis MacDonald Cornford. Oxford: Oxford University Press, 1945.

Pocock, John G. A. *The Machiavellian Moment: Florentine Political Thought and the Atlantic Republican Tradition*. Princeton, NJ: Princeton University Press, 1975.

Popper, Karl R. *The Open Society and Its Enemies*. Two volumes. Princeton: Princeton University Press, 1962.

Porter, J. M. "Rousseau: Will and Politics." In *Unity Plurality and Politics*. Edited by J. M. Porter and Richard Vernon. New York: St. Martin's Press, 1986.

Postman, Neil. *Technopoly: The Surrender of Culture to Technology*. New York: Alfred A. Knopf, 1992.

Price, Vincent. *Public Opinion*. Newbury Park, CA: Sage Publications, 1992.

Przeworski, Adam, Susan Stokes, and Bernard Manin. *Democracy, Accountability, and Representation*. Cambridge: Cambridge University Press, 1999.

Putnam, Hilary. *Renewing Philosophy*. Cambridge, MA: Harvard University Press, 1992.

Putnam, Robert, D. *Bowling Alone: The Collapse and Revival of American Community*. New York: Simon and Schuster, 2000.

Radcliff, Benjamin. "The General Will and Social Choice Theory." *The Review of Politics* (Winter 1992): 34-49.

Raphael, Ray. *A People's History of the American Revolution*. New York: The New Press, 2001.

Ravitch, Diane, and Abigail Thernstrom, eds. *The Democracy Reader*. New York: Harper Collins Publishers, 1992.

Rawls, John. *A Theory of Justice*. Cambridge, MA: The Belknap Press of Harvard University, 1971.

——. "Justice as Fairness: Political Not Metaphysical." *Philosophy and Public Affairs*, vol.14, no.3, (Summer 1985).

——. *Justice as Fairness: A Brief Restatement*. Cambridge, MA: Harvard University Press, 1989.

——. *Political Liberalism*. New York: Columbia University Press, 1993.

——. *The Law of Peoples*. Cambridge, MA: Harvard University, 1999.

Raz, Joseph. *The Morality of Freedom*. Oxford: Clarendon Press, 1986.

Riker, William H. *Liberalism Against Populism*. Prospect Heights, IL: Waveland Press, Inc., 1982.

Reynolds, James F., and David C. Paris. " The Concept of 'Choice' and Arrow's Theorem." *Ethics*, vol.89, no.4 (July 1979): 354-71.

Rheingold, Howard. *The Virtual Community: Homesteading on the Electronic Frontier*. New York: Addison-Wesley Publishing Company, 1993.

Risse, Mathias. "Arrow's Theorem, Indeterminacy, and Multiplicity Reconsidered." *Ethics*, vol.111 (July 2001): 706-34.

Rochlin, Gene I. *Trapped in the Net: The Unanticipated Consequences of Computerization*. Princeton: Princeton University Press, 1997.

Roemer, John. *A Future for Socialism*. Cambridge, MA: Harvard University Press, 1994.

Rorty, Richard. "The Priority of Democracy to Philosophy." In *Prospects for a Common Morality*. Edited by Gene Outka and John P. Reed. Princeton: Princeton University Press, 1993.

Rosenthal, Alan. *The Decline of Representative Democracy: Process, Participation, and Power in State Legislatures*. Washington, D.C.: Congressional Quarterly, 1998.

Rousseau, Jean-Jacques. *On the Social Contract* (1762). Edited and translated by Donald A. Cress, introduced by Peter Gay. Indianapolis: Hackett Publishing Company, 1987.

——. *Emile* (1762). Translated by Barbara Foxley, introduced by P. D. Jimack. London: Everyman, 1993.
——. *Discourse on the Origins of Inequality* (1762). Translated by Donald A. Cress, introduced by James Miller. Indianapolis, IN: Hackett Publishing Company, 1992.
——. *Political Writings*. Translated and edited by Frederick Watkins. Madison: The University of Wisconsin Press, 1986.
Rubel, Maximilien. "Marx's Concept of Democracy." *Democracy*, vol.3, no.4 (Fall 1983).
Ryden, David K. *Representation in Crisis: The Constitution, Interest Groups, and Political Parties*. Albany: State University of New York Press, 1996.
Sandel, Michael. *Liberalism and Limits of Justice*. Cambridge: Cambridge University Press, 1982.
——. "The Procedural Republic and the Unencumbered Self." *Political Theory*, vol.12, no.1 (February 1984).
——. *Democracy's Discontent*. Cambridge: Belknap Press of Harvard University Press, 1996.
Sarri, Donald G. *Basic Geometry of Voting*. Berlin: Springer, 1995.
——. *Decisions and Elections: Explaining the Unexpected*. Cambridge: Cambridge University Press, 2001.
Saward, Michael. *Democratic Innovation: Deliberation, Representation, and Association*. London: Routledge, 2000.
Schickler, Eric. "Democratizing Technology: Hierarchy and Innovation in Public Life." *Polity*, vol.xxvii, no.2 (Winter 1994): 175-99.
Schmitt, Carl. *The Crisis of Parliamentary Democracy* (1923). Translated by Ellen Kennedy. Cambridge, MA: The MIT Press, 1988.
Schram, Martin. *Speaking Freely: Former Members of Congress Talk about Money in Politics*. Washington, D.C.: Center for Responsive Politics, 1995.
Schumpeter, Joseph. *Capitalism, Socialism, and Democracy*. New York: Harper Colophon Books, 1942.
Schwartz, David B. *Who Cares? Rediscovering Community*. Oxford: Westview Press, 1997.
Sclove, Richard E. *Democracy and Technology*. New York: The Guilford Press, 1995.
Selznick, Philip. *The Moral Commonwealth: Social Theory and the Promise of Community*. Berkeley: University of California Press, 1992.
Sen, Amartya K. "Rational Fools: A Critique of the Behavioral Foundations of Economic Theory." *Philosophy and Public Affairs*, vol.6, no.1 (1977): 317-44.
——. *Welfare Economics and the Real World*. Memphis: P. K. Seidman Foundation, 1986.
——. "Rationality and Social Choice." *The American Economic Review*, vol.85, no.1 (March 1995):1-24.
——. "Individual Preference as the Basis of Social Choice." In *Social Choice Re-Examined: Proceedings of the IEA Conference Held at Schloss Hern-*

stein, Berndorf, Vienna, Austria. Edited by Kenneth Arrow, Amartya Sen, and Kotaro Suzumua, vol.1. London: Macmillan Press LTD, 1996.
Simon, Roger. *Gramsci's Political Thought: An Introduction*. London: Lawrence and Wishart, 1982.
Sinclair, R. K. *Democracy and Participation in Athens*. Cambridge: Cambridge University Press, 1988.
Sirianni, Carmen. *Workers Control and Socialist Democracy: The Soviet Experience*. London: Verso New Left Books, 1982.
Skinner, Quentin. "On Justice, the Common Good and the Priority of Liberty." In *Dimensions of Radical Democracy: Pluralism, Citizenship, Community*. Edited by Chantal Mouffe. London: Verso, 1992.
——. "The Italian City-Republics." In *Democracy the Unfinished Journey: 508 BC to AD 1993*. Edited by John Dunn. Oxford: Oxford University Press, 1992b.
——. *Liberty Before Liberalism*. Cambridge: Cambridge University Press, 1998.
Slaton, Christa Daryl. *Televote: Expanding Citizen Participation in the Quantum Age*. New York: Praeger, 1992.
Slonim, S., ed. *Framers' Constitution/Beardian Deconstruction: Essays on the Constitutional Design of 1787*. New York: Peter Lang Publishing, 2001.
Slouka, Mark. *War of the Worlds: Cyberspace and the High-Tech Assault on Reality*. New York: Basic Books, 1995.
Smart, Paul. *Mill and Marx: Individual Liberty and the Roads to Freedom*. Manchester: Manchester University Press, 1991.
Snider, James H. "Democracy On-Line." *The Futurist* (September-October 1994): 15-19.
Stalley, R. F. *An Introduction to Plato's Laws*. Indianapolis: Hackett Publishing Company, 1983.
Sugden, Robert. *The Political Economy of Public Choice*. New York: John Wiley & Sons, 1981.
Sunstein, Cass R. "Preferences and Politics." *Philosophy and Public Affairs*, vol.20, no.1 (Winter 1991): 3-34.
——. *Republic.Com*. Princeton: Princeton University Press, 2001a.
——. *Designing Democracy: What Constitutions Do*. Oxford University Press, 2001b.
Tangian, A. S. "Unlikelihood of Condorcet's Paradox in a Large Society." *Social Choice and Welfare*, no.17 (2000): 337-65.
Thompson, Dennis F. *John Stuart Mill and Representative Government*. Princeton: Princeton University Press, 1976.
Thorley, John. *Athenian Democracy*. London: Routledge, 1996.
Tocqueville, Alexis de. *Democracy in America*. In 1 vol. (1835 and 1840). Edited by Harvey C Mansfield and Delba Winthrop. Chicago: The University of Chicago Press, 2000.

Townshend, Jules. "C. B. Macpherson: Capitalism, Human Nature and Contemporary Democratic Theory." In *Marxism's Ethical Thinkers*. Edited by Lawrence Wilde. New York: Palgrave, 2001

Trachtenberg, Zev M. *Making Citizens: Rousseau's Political Theory of Culture*. London: Routledge, 1993.

Trotsky, Leon. *Literature and Revolution*. New York: Russell & Russell, 1957.

——. *Terrorism and Communism*. Ann Arbor: The University of Michigan Press, 1961.

——. *The Revolution Betrayed*. New York: Pathfinder Press, 1972.

——. *Problems of Everyday Life*. New York: Monad Press, 1973.

Tucker, Robert. *Philosophy and Myth in Karl Marx*. Cambridge: Cambridge University Press, 1961.

——. "Lenin's Bolshevism as a Culture in the Making." In *Bolshevik Culture*. Edited by A. Gleason, P. Kenez, and R. Stites. Bloomington: Indiana University Press, 1985.

Urbinati, Nadia. "Detecting Democratic Modernity: Antonio Gramsci on Individuals and Equality." *The Philosophical Form* xxix, no.3-4 (Spring-Summer 1998): 168-81.

——. "Democracy and Populism." *Constellations,* vol.5, no.1 (1998): 110-24.

Valadez, Jorge M. *Deliberative Democracy, Political Legitimacy, and Self-Determination in Multicultural Societies*. Boulder, CO: Westview Press, 2001.

Vernon, Richard. *Political Morality*. New York: Continuum, 2001.

Viale, Riccardo, ed. *Knowledge and Politics*. New York: Physica-verlag, 2001.

Wainwright, Hilary. *Reclaim the State: Experiments in Popular Democracy*. London: Verso, 2003.

Waldron, Jeremy. *Liberal Rights: Collected Papers 1981-1991*. Cambridge: Cambridge University Press, 1993.

Walzer, Michael. "Philosophy and Democracy." *Political Theory,* vol.9, no.3 (August 1981): 379-99.

Weale, Albert. *Democracy*. New York: St. Martin's Press, 1999.

Weare, Christopher, Juliet A. Musso, and Mathew L. Hale. "Electronic Democracy and the Diffusion of Municipal Web Pages in California." *Administration & Society,* vol.31, no.1 (March 1999): 3-27.

Welch, Cheryl B. *De Tocqueville*. Oxford: Oxford University Press, 2001.

Wellen, Richard. *Dilemmas in Liberal Democratic Thought Since Max Weber*. New York: Peter Lang, 1996.

West, Darrell M. *Checkbook Democracy: How Money Corrupts Political Campaigns*. Boston: Northeastern University Press, 2000.

Wildavsky, Aaron. "Choosing Preferences by Constructing Institutions: A Cultural Theory of Preference Formation." *American Political Science Review,* vol.81, no.1 (March 1987).

Wilhelm, Anthony G. *Democracy in a Digital Age: Challenges to Political Life in Cyberspace*. New York: Routledge, 2000.

Williams, Beryl. *Lenin*. London: Longman, 2000

Williams, Melissa H. *Voice, Trust, and Memory*. Princeton: Princeton University Press, 1998.

Wingrove, Elizabeth Rose. *Rousseau's Republican Romance*. Princeton, NJ: Princeton University Press, 2000.

Winner, Langdon. *The Whale and the Reactor*. Chicago: The University of Chicago Press, 1986.

——, ed. *Democracy in a Technological Society*. Boston: Kluwer Academic Publishers, 1992.

Wolfensberger, Donald R. *Congress and the People: Deliberative Democracy on Trial*. Baltimore: Johns Hopkins University Press, 2000.

Wolff, Robert Paul. *In Defense of Anarchism*. New York: Harper Torch Books, 1970.

Wolin, Sheldon S. *Politics and Vision: Continuity and Innovation in Western Political Thought*. Boston: Little, Brown and Company, 1960.

Wood, Allen. *Karl Marx*. London: Routledge & Kegan Paul, 1981.

Wood, Donald N. *The Unraveling of the West: The Rise of Postmodernism and the Decline of Democracy*. Westport, CT: Praeger, 2003.

Wood, Gordon S. "The Enemy Is Us: Democratic Capitalism in the Early Republic." In *Wages of Independence: Capitalism in the Early American Republic*. Edited by Paul A. Gilje. Madison: Madison House, 1997.

——. *The Radicalism of the American Revolution*. New York: Alfred A. Knopf, 1992a.

——. "Democracy and American Revolution." In *Democracy the Unfinished Journey: 508 BC to AD 1993*. Edited by John Dunn. Oxford: Oxford University Press, 1992b.

——. *The Creation of the American Republic 1776-1787*. Chapel Hill: The University of North Carolina Press, 1969.

Woolley, Peter, and Albert R. Papa. *American Politics: Core Arguments/Current Controversy*. Upper Saddle River: NJ: Prentice Hall, 2002.

Wootton, David. "The Levellers." In *Democracy the Unfinished Journey: 508 BC to AD 1993*. Edited by John Dunn. Oxford: Oxford University Press, 1992.

Xenos, Nicholas. "Democracy as Method: Joseph A. Schumpeter." *Democracy*, vol.1 (1981): 110-23.

Yankelovich, Daniel. *Coming to Public Judgement: Making Democracy Work in a Complex World*. Syracuse: Syracuse University Press, 1991.

Young, Alfred F. *Beyond the American Revolution: Explorations in the History of American Radicalism*. DeKalb, IL: Northern Illinois University Press, 1993.

Young, Iris Marion. "Social Groups in Associative Democracy." In *Associations and Democracy: The Utopian Project*, vol.1. Edited by Joshua Cohen and Joel Rogers. London: Verso, 1995.

Zinn, Howard. *A People's History of the United States*. New York: Harper Colophon Books, 1980.

Index

About the Author

Majid Behrouzi holds a Ph.D. in philosophy from York University (Canada). In addition to philosophy, he has also studied mathematics and engineering and holds advanced degrees in these fields. He currently teaches mathematics and philosophy in Cleveland.